Robotics

This volume is one of a series that examines
various aspects of computer technology
and the role computers play in modern life.

UNDERSTANDING COMPUTERS

Robotics

BY THE EDITORS OF TIME-LIFE BOOKS
TIME-LIFE BOOKS, ALEXANDRIA, VIRGINIA

Contents

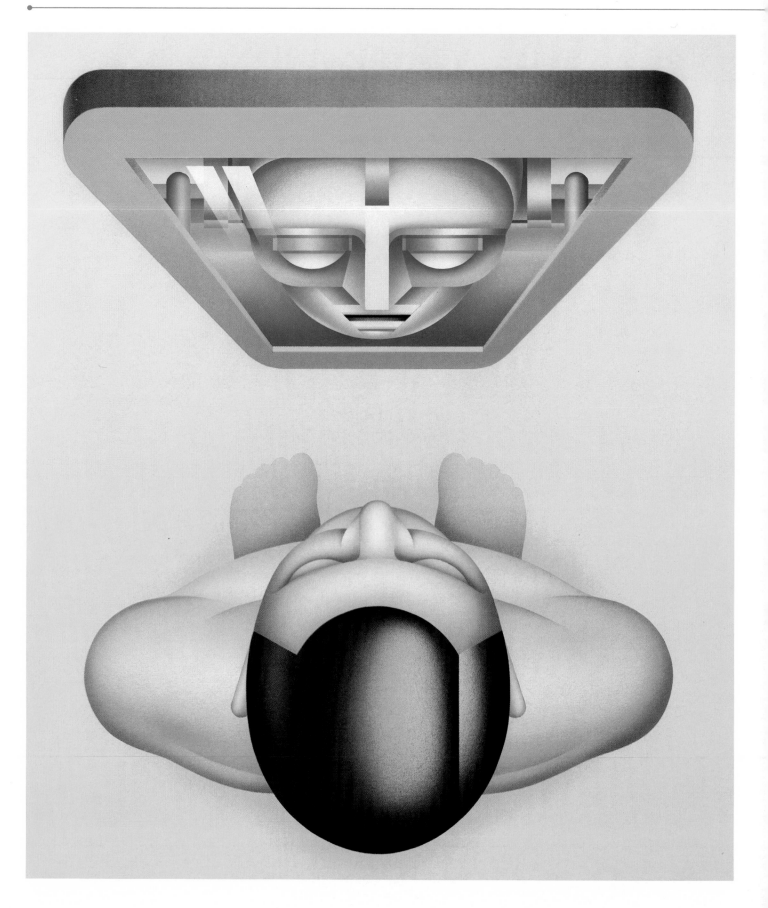

6

The Search for Surrogates

Hurtling through space at more than 60,000 miles per hour, an explorer named *Voyager 2* in 1989 ventured where none had gone before. Its mission: to give earthbound scientists a close-up of the planet Neptune. The probe was scheduled to swoop within 3,000 miles of the planet's blue cloud cover on August 24. Into a scant time-window of about ten hours, mission controllers at NASA's Jet Propulsion Laboratory (JPL) in Pasadena, California, hoped to cram a host of critical scientific observations — all of which *Voyager* would have to perform without a single direct command from earth. The craft was certainly capable of receiving radio signals from JPL: Its 12-foot-wide dish antenna was designed to point homeward for just this reason. But signals traveling in either direction would take more than four hours to cover the intervening 2.7 billion miles. For purposes of the important observations, then, *Voyager* had to be programmed not only to perform specific tasks but to react on its own to a set of anticipated problems; real-time reaction from earth was impossible.

Launched in 1977, *Voyager* had been serving beyond its original specifications for eight years. The craft's six onboard computers and the battery of sensors and cameras arrayed on its spiky instrument booms had been designed for only five years of intensive space exploration — to survey Jupiter in 1979 and Saturn in 1981. But before *Voyager* had passed Saturn, NASA decided to extend the mission: The probe would go to Uranus and from there to Neptune, using the gravitational pull of Saturn and then of Uranus to deflect it onto the proper trajectory for its next assignment. The change in plan meant that *Voyager's* handlers had to reconfigure much of the probe's controlling hardware and software.

Over the next few years, as *Voyager* continued its grand tour of the solar system, a team of flight engineers radioed instructions to prepare the spacecraft for the special challenges it would meet. For example, because Neptune is about 30 times farther from the sun than the earth is, and one and a half times as far as Uranus, *Voyager's* cameras would have to take extremely long exposures to overcome the scarcity of light, a constraint certain to produce only blurry images. Thus, among *Voyager's* new instructions were some commanding it to rotate slowly to compensate, so that its cameras could follow their target in much the same way that a human photographer pans the camera to capture a speeding object. The probe first applied the new picture-taking method at Uranus in 1986. The revised software proved its worth, returning flawless images of eleven rings and fifteen moons, ten of them previously unknown.

Still ahead for *Voyager* lay the three and a half year journey to Neptune, a world so distant that scientists could only speculate about its surface features and other attributes. During the probe's passage to Neptune, JPL controllers monitored transmissions from the spacecraft and occasionally sent out new instructions to correct minor malfunctions. These ministrations from earth paid off handsomely. Speeding past Neptune, *Voyager* beamed back a flood of data describing, among other things, the contents of the planet's atmosphere and the strength and orientation of its magnetic field. Moreover, the imaging software tested at Uranus revealed previously unknown features of Neptune and its environs: a huge "dark spot" on the surface; three rings girdling the planet; and six moons, raising the known complement to eight. Astronomers, physicists and geologists learned as much about Neptune from *Voyager's* fleeting survey as they had in decades of telescopic study from earth.

A BROAD DEFINITION

A scientific probe arrowing through the solar system may not seem a likely candidate for the appellation "robot." But in its performance of programmed — and especially reprogrammed — tasks, *Voyager 2* more than earns the name. Acting for long periods without direct human supervision, the probe has traveled to places not yet accessible to astronauts made of flesh and blood. And en route, it has served admirably as a surrogate for human explorers.

On earth as well, robots in various forms carry out an assortment of tasks for their human masters. They have sheared sheep in Australia and formed rice cakes for *sushi* in Japan (where most of the world's population of robots resides). In several European cities, robotic trains carry commuters to and from work, and in one American hospital, a robot arm locates brain tumors with an accuracy no human surgeon could hope to match. On the battlefront, so-called smart weapons can navigate across hundreds of miles to their targets *(pages 26-27)*. And on assembly lines in many parts of the industrial world, the most numerous members of the robot family labor tirelessly *(pages 47-59)*: Arranged in efficient groups on the factory floor, they spot-weld and spray-paint car bodies, machine parts for fighter planes and even assemble delicate watches.

What all these machines — from *Voyager 2* to a spray-painting industrial robot — have in common is that they perform tasks too tedious, too precise or too dangerous for fallible and mortal human beings. In space, at the bottom of the ocean or in a paint booth engulfed in toxic fumes, robotic machines function without benefit of oxygen and can withstand extremes of temperature. Provided they are properly programmed and maintained, robots in industry will repeat a task correctly and precisely hundreds of thousands of times, and they will never grow bored or careless.

Among scientists and others in the field, the definition of a robot is subject to sometimes intense debate, and no definition has gained universal acceptance. Members of the Robotic Industries Association (RIA) — founded in 1974 as the Robot Institute of America and made up primarily of manufacturers and users of industrial robots — wrangled long and hard before hammering out a rather wordy compromise definition in 1979. By that edict, a robot was "a reprogrammable, multifunctional manipulator designed to move material, parts, tools or specialized devices through variable programmed motions for the performance of a variety of tasks." Less than a decade later, the compromise had to be rewritten to include, among other things, machines equipped with vision and other sensory systems. Outside industry, researchers in laboratories around the world work on machines with skills other than manipulation — legged locomotion, for example, or autonomous navigation.

In any case, the original RIA definition, with its industrial assembly-line emphasis, did not embrace a number of machines that employ robotic technology. Strictly speaking, the *Voyager* probe is not a manipulator. And remotely controlled devices sometimes known as teleoperators *(pages 115-121)* are directed from a distance by humans rather than solely by computer program. At the opposite end of the spectrum are widely used but fairly unsophisticated machines called pick-and-place, or "bang-bang," robots. Although these devices, which make up a large part of what the Japanese include in their own robot count, are programmed to move materials or parts, their range of movement is ultimately determined less by computer program than by mechanical stops — fixed blocks placed at either end of each of the robot arm's axes of movement (hence the bang-bang). But regardless of what the professionals decree, a robot in the popular imagination is a humanoid machine endowed, for better or worse, with superhuman strength or brain power.

The word itself was coined in the 1920s by playwright Karel Čapek from the Czech *robota*, meaning "work." In Čapek's play *R.U.R.* (for Rossum's Universal Robots), human-like mechanical creatures produced in Rossum's factory are docile slaves until one misguided scientist gives them emotions. The robots revolt, kill all humans and take over the world. Čapek wrote *R.U.R.* just after World War I, when machines — in the form of tanks and other mechanized weapons — seemed to bode more ill than good for humanity. Two decades later, the prolific science fiction writer Isaac Asimov promoted a more appealing vision. In a series of stories and novels, Asimov (who invented the word "robotics" to refer to the science of robots) imagined a world in which mechanical beings were humanity's devoted helpmates. They were constrained to obey three interrelated Laws of Robotics: 1. A robot may not injure a human being or, through inaction, allow a human being to come to harm. 2. A robot must obey the orders given it by human beings except where such orders would conflict with the First Law. 3. A robot must protect its own existence as long as such protection does not conflict with the First or Second Law.

THE RISE OF AUTOMATIC CONTROL

By 1940, more and more real-world machines seemed to be running themselves. In the United States, magazine articles hailed a number of devices as "mechanical men" or "thinking machines." Included in this category were punched-card tabulators and automatic door openers, as well as more elaborate systems for inspecting or controlling industrial processes. In sheet-steel plants, for example, photoelectric cells measured lengths of steel for cutting. In textile mills they counted crosswise threads, zipping by as fast as 10,000 threads a second, to keep bolts of cloth miles long from skewing on the rollers. In the maws of giant stamping machines, photocells acted as a safeguard, preventing the machine from closing if the operator's hand was in the way. Where machines had once replaced only human muscle, they now seemed to be replacing the human brain. In truth, however, the evolution of robots from fantasy to reality was just beginning. World War II, with its demands for weaponry, would be the catalyst needed to advance theory and technology enough to bring real robots into being.

In the spring of 1940, American newspapers and radio broadcasts were full of the disheartening news from Europe: Hitler's armies were rapidly advancing through Holland, Belgium and France, largely because of Germany's strength in the air. The Allies were desperate for faster and more accurate methods of directing antiaircraft fire, both on land and at sea.

In late May, as Stuka bombers rained terror on ships attempting to evacuate British troops from the beaches at Dunkirk, a 29-year-old engineer named David B. Parkinson was working with a small group at Bell Telephone Laboratories. The team was trying to improve the accuracy of automatic level recorders, devices used to measure voltage in telephone transmission circuits. In the group's new design, a potentiometer — a type of mechanically variable resistor — regulated the movement of a pen across the width of a strip of paper that was traveling lengthwise at a uniform speed. An inch and three quarters in diameter, the potentiometer consisted of a specially shaped card with a length of resistance wire wound around it. When a rapidly varying voltage, such as that of a telephone transmission, was applied across the terminals of the potentiometer, it was accurately reflected by the pen's proportional movement.

A DREAM CONNECTION

One night in early June, after Parkinson had been working on the level recorder for several weeks, he had, as he wrote many years later, "the most vivid and peculiar dream." In the dream, he found himself behind an embankment with an antiaircraft gun crew. "The men were Dutch or Belgian by their uniforms," he remembered. Parkinson, a physicist by training, possessed only the most general knowledge of artillery, but he knew enough to realize that something unusual was going on: The gun was firing occasionally, but "every shot brought down an airplane!" After three or four shots, one of the men in the crew smiled at him and beckoned him closer, pointing to one of the pivots on which the gun swiveled. "Mounted there was the control potentiometer of my level recorder," Parkinson wrote. "There was no mistaking it — it was the identical item."

When he awoke, the import of the dream was clear: "If the potentiometer could control the high-speed motion of a recording pen with great accuracy, why couldn't a suitably engineered device do the same thing for an antiaircraft gun!" Told of the dream, Parkinson's supervisor, Clarence A. Lovell, was infected by his friend's excitement. As Parkinson recalled, "We both pitched head over heels into a study of the possibilities." Neither had any experience with fire control, but when they presented the idea to management, Bell Labs recognized that the army should be told immediately.

At the time, gun directors, as the devices that controlled antiaircraft artillery were called, were essentially mechanical analog computers, performing mathematical calculations through the movements of gears, shafts and other machined components. The calculations involved the complex differential equations needed to predict — however roughly — where a moving target would be a few moments in the future; more computations then determined what the gun's elevation, range and direction settings should be, as well as the appropriate setting for a time fuse to cause the shell to burst as close to the target as possible.

Although the mechanical director could solve the necessary equations faster than a human could, it relied on the human gun crew, using manually operated optical tracking equipment, to supply the information it needed about the target's most recent positions and speed. As often as not, optical tracking was hampered—if not made impossible—by darkness, clouds, fog or the smoke of battle. Even if visibility was good and the necessary equations were solved, the settings had to be transmitted to an indicator on the gun itself. Then a human gunner turned wheels and rotated gears to bring the weapon into compliance with the gun director's orders. Other men, meanwhile, set the time fuse and prepared to load the shell into the gun. All in all, this method of prediction and execution, adequate for relatively slow-moving targets such as ships or for stationary ones such as enemy gun emplacements, was far too slow to be effective against aircraft.

By 1940, optical tracking was on its way to being replaced by the new technology of radar, providing an opportunity to improve the entire system of antiaircraft fire control. With radar tracking, the electromagnetic pulses reflected by an aircraft caught in a radar beam could supply a continuous stream of information about the target's position, speed and direction. If the gun director were also electrical—and if the gun itself could be controlled electrically—it would be possible to make split-second adjustments in the gun's aim.

After a series of meetings with the newly formed National Defense Research Committee (NDRC), Bell Labs and Bell's manufacturing division, Western Electric, were awarded the contract to develop and build an electrical analog director to control the U.S. Army's 90-millimeter antiaircraft gun. Over the next two years, the scientists worked not only on the director itself but on the many mechanisms that would enable commands from the director to be carried out by the gun without human intervention.

Just before Christmas, 1942, the first production-model director, designated the M-9, was delivered to the army; early the following year, it went into action. Linked to the penetrating eye of a radar system, the M-9 could calculate a target's probable future position and cause the massive 90-millimeter gun to take aim automatically, much as a modern-day computer can direct the actions of an industrial robot arm. All that the gun's human tenders had to do was keep the weapon loaded. The M-9 acquitted itself well on several fronts, but perhaps its most remarkable performance came in August 1944, during the Second Battle of Britain. Together with new "proximity" fuses—vacuum tubes built into the nose of the explosive shell and designed to detonate on approaching the target—the M-9 was responsible that month for downing nine out of 10 German V-1 buzz bombs headed for London.

CLOSING THE CONTROL LOOP
Proximity fuses, radar and electrical computing gear were all critical to the new generation of antiaircraft weapons. But what allowed these disparate elements to form a coordinated and deadly system was the rapid development of so-called servomechanism theory. Servomechanisms (from the Latin *servus,* or "slave") are devices that take information from one stage of a process and feed it back to a stage earlier in the cycle to drive the system into compliance with a commanded value. In an antiaircraft gun system, servomechanisms made up of potentiometers, feedback amplifiers and motors performed that function at various places in the loop. The potentiometers—larger versions of the one in David Parkinson's

A Closed Loop for Precise Execution

To execute instructions accurately, most robots rely on feedback in the form of a continuous stream of data. In very sophisticated systems, most of which are still in the experimental stage, this data can come from a variety of outward-directed devices, such as vision systems *(pages 19-31)* or tactile sensors *(pages 90-99)*. In most industrial robots, it is generated by devices that, in effect, look inward to measure the positions and the rate of movement of the robot's joints *(page 57)*. (Controlling a robot through feedback is called closed-loop control. Some rudimentary robots, known as open-loop types, incorporate no feedback; they depend instead on mechanical stops to control movement.)

In essence, a feedback loop consists of a mechanism and the environment upon which it acts. A household thermostat is a simple feedback loop: Within the thermostat, a bimetallic strip changes shape in reaction to temperature and thereby turns a heating or cooling unit on or off. As illustrated here, a robotic feedback loop adds a third element — a computerized controller. The loop begins when the controller issues a command, usually in digital form, to the robotic mechanism. The mechanism responds with a physical action that modifies either its internal or external environment — rotating a joint, for example, or arc-welding a metal seam. This modification is detected and measured by sensors, which report the precise effects to the controller. The controller calculates the difference between the actual and desired results and closes the loop by issuing a corrective command to the mechanism.

Unlike a simple thermostat, the controller can be programmed to react in various ways to the feedback. For example, depending on the job, it may disregard small discrepancies between desired and actual results, but report to a human supervisor if it detects a trend toward greater and greater errors.

Feedback

CONTROLLER

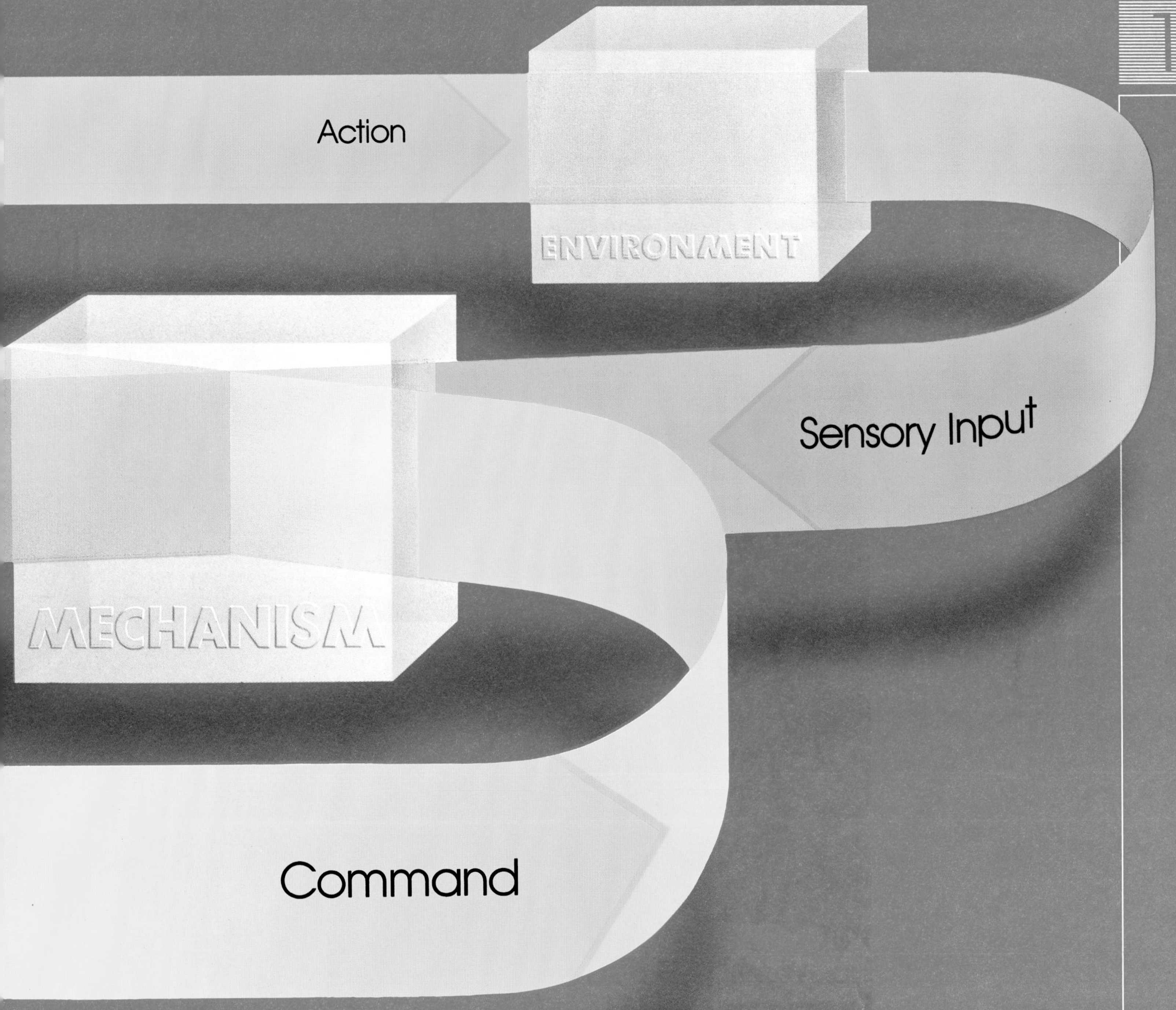

Action

ENVIRONMENT

Sensory Input

MECHANISM

Command

dream — in effect compared the gun's actual movement (measured as shaft rotations and converted to voltage) with the movement dictated by the gun director (raising the gun to a certain elevation, say), and corrected control voltages as needed to drive motors that turned the appropriate shafts. With each correction, the cycle began again, to be continuously and rapidly repeated until the gun was properly positioned.

The feedback cycle of measure, compare, correct and remeasure had long been known to engineers in a general fashion, but wartime research brought a much deeper level of understanding. One of the most active centers of this research was the Massachusetts Institute of Technology, which had several groups investigating different aspects of the problem of antiaircraft fire control. The M.I.T. Radiation Laboratory, for instance, designed a radar system, the SCR 584, with output potentiometers specifically suited to link it with Bell Labs' M-9 gun director. And the aptly named Servomechanisms Laboratory, organized in late 1940 by the school's electrical engineering department, developed remote-control systems for 40-millimeter gun drives and for both airborne and naval radar and gun equipment.

The Servomechanisms Lab was headed by Professor Gordon S. Brown, who fostered an atmosphere of free inquiry rare for that day. What he valued most about the wartime scientific enterprise was the opportunity for researchers in different fields to pool abilities. "Specialists of engineering and science found themselves talking to one another for the first time in generations," Brown remembered later. "Mechanical engineers exploited techniques of circuit theory borrowed from the communications engineers; mathematicians working with engineers and experimental scientists discovered entirely unsuspected practical uses for forgotten theorems." Out of this enforced collaboration came many of the techniques that would eventually be put to use by roboticists interested in designing hardworking robots for industry. Mathematicians supplied the theoretical underpinnings for control systems that engineers could realize through ingenious arrangements of motors, amplifiers and hydraulic transmissions. "It became possible to design control-system components with entirely new properties," Brown wrote in 1952.

A PRODIGY'S WORK
One mathematician who instinctively favored the collaborative approach Brown espoused was M.I.T. professor Norbert Wiener, a genius whose wide-ranging inquiries embraced biology, neurophysiology, electrical engineering and the nascent field of digital computing. Out of this eclectic assortment of interests, Wiener would fashion new connections between human and machine. Nearly half a century later, the influence of his ideas can still be perceived in research carried out at the frontiers of robotics (Chapter 3).

Born in 1894 in Columbia, Missouri, Wiener grew up in New England, the son of a professor of Slavic languages at Harvard who demanded extraordinary intellectual achievement from all of his children. The boy Norbert could read by the age of three; by eight, threatened with myopia, he had to stop reading for six months and listen to oral lessons given by his father. Wiener entered Tufts College at 11. In 1913, at 18, he obtained his Ph.D. from Harvard with a dissertation on the philosophy of mathematics. There followed a few years of postgraduate work, a couple of hack writing jobs and a brief stint as a human computer at the U.S. Army's Aberdeen Proving Grounds in Maryland, where he worked on the equations needed to compute ballistic firing tables. Then, in 1919, Wiener joined the mathematics department at M.I.T. He took the position a bit reluctantly, for at the time the department was small and undistinguished, intended merely to teach mathematics to engineering students. Over the next four decades, however, he and the department would grow in professional stature together.

In 1940, Vannevar Bush, former dean of M.I.T.'s School of Engineering and newly appointed chairman of the National Defense Research Committee, asked his colleagues for ideas on how best to mobilize the scientific community if the United States entered the war. Wiener had two suggestions, neither of which Bush took him up on, but both reflecting Wiener's penchant and aptitude for cross-disciplinary work. First, he supported the formation of small teams of scientists from different fields to attack given problems in concert. The association should be voluntary, he felt, to preserve "a large measure of the scientists' initiative and individual responsibility."

THE WIENER TREATMENT
Wiener's commitment to interdisciplinary collaboration, genuine as it was, was often derailed by a competitiveness that masked recurrent feelings of inferiority. Short, nearsighted and rotund, he was known to fall promptly asleep during speeches by others — his "habitual defense against competition," according to a British neurologist who received the Wiener treatment on their first meeting in 1946. Later, after the two had become friends, the neurologist recorded Wiener's brain rhythms and found that in fact "his defensive naps were real deep sleep; he could drop off in a few seconds but would waken instantly if one spoke his name or mentioned a topic in which he was really interested."

Wiener's second suggestion was that Bush direct research efforts toward the design of electronic digital computers for solving partial differential equations of the kind essential to accurate antiaircraft fire control, among other things. In the several years preceding the outbreak of war, Wiener had done considerable work on electrical circuits, even investigating, together with an engineering colleague in China, the possibility of speeding up analog computers by using electrical instead of mechanical relays. As one historian noted, however, "Their inadequate understanding of feedback mechanisms at that time precluded success."

Bush himself had a longstanding interest in analog computing machines. In 1930, he had built the differential analyzer, a mechanical analog computer that was the first machine capable of solving complex differential equations, and one of his analyzers was even then being used to compute firing tables at the Aberdeen Proving Grounds. But believing that the technology to build a digital machine was not advanced enough to bear fruit in time for the war, Bush recommended that Wiener hold off on this idea for the duration.

Deprived of what he had hoped would be a wartime project with longer-range consequences, Wiener turned almost by default to the problem that many of his colleagues were working on — namely, improving the design of control systems for antiaircraft guns. But while other researchers looked to make better servo systems for translating the gun director's commands into action, Wiener was drawn to the mathematical challenge of improving the way the director arrived at those commands in the first place. The problem was tailor-made for him, allowing him to integrate a number of his interests, not only in mathematics and engineering but in human physiology as well. As it happened, the work would lead in an unexpected direction.

"It was necessary to build into the control system of the antiaircraft gun some mechanical equivalent of a range table," Wiener wrote later, "which would automatically allow the gun the necessary lead over the plane to make the shell and the plane come to the same place at the same time. To some extent this is a purely geometrical problem, but in its finer developments it involves an improvement of our estimate of the future position of the plane itself." The "finer developments" included the essentially unpredictable evasive action sure to be taken by the pilot.

Wiener set to work with vigor but soon hit a snag. "The mathematical processes which suggested themselves to me in the first instance for prediction were, in fact, impossible of execution," he wrote, "for they assumed an already existing knowledge of the future." But then he arrived at a statistical method that seemed to allow him to approximate a prediction. At this point, his ideas looked so promising that they were immediately classified by the NDRC, and he was assigned an assistant: Julian Bigelow, a young engineer from IBM.

A COMPLEMENTARY SKILL

It was the beginning of what would be a long and fruitful collaboration. Wiener once described Bigelow as "a quiet, thorough New Englander, whose only scientific vice is an excess of scientific virtue. He is a perfectionist, and no work that he has ever done is complete enough in his eyes to satisfy him." Bigelow was also the collector of, and expert mechanical doctor to, a series of decrepit automobiles that, "by all the canons of the motorist," according to Wiener, "should have been consigned to the junk heap years ago." Bigelow's skill with machinery was the ideal complement to Wiener's mathematical prowess.

Over a period of several months, Wiener and Bigelow struggled with the problem of incorporating the human elements of gunner and pilot into the mechanism of the gun control system as a whole. When they began their work, radar had not yet come into its own; later, the tracking of an aircraft was automated, virtually eliminating the need to consider the gunner's responses in the overall control system. But it was not possible to eliminate the effect of evasive responses by the human pilot. To produce a gun director based on a sound mathematical treatment of the entire control problem, Wiener and Bigelow looked for mechanical analogies of gunner and pilot.

They built two experimental devices for the purpose. One machine, controlled by the statistically based equations that Wiener had previously arrived at, modeled the irregular behavior of an aircraft under fire by creating a patch of light that danced about on the ceiling like the shadow of a fly. With the other machine — representing the gunner — a person could pursue the patch by turning a crank to manipulate a mirror that reflected a separate spot of light. Wiener and Bigelow deliberately made the second apparatus complicated and unnaturally difficult to control, so that each person who tried it would respond differently. Thus, the two researchers could factor in not only the general behavior of the machine but the ability of an average individual.

What they found was that the human operators seemed to regulate their actions by observing the errors of one pattern of behavior — cranking hard to the right, say — and then countering the errors with action intended to reduce them. This method of control — often called negative feedback — was, of course, already being employed in electric circuits and various types of servo systems. In an effort to gain a clearer understanding of negative feedback as applied to humans, Wiener and Bigelow followed a maxim of physiologists — namely, that one can learn about the normal behavior of an organ or a system by studying its pathological behavior. In servomechanisms, for instance, an aberration called hunting was one of the possible dangers of feedback control systems: Excessive feedback causes the machine to overshoot its mark repeatedly and to fluctuate wildly back and forth, without settling on the desired value. Wiener and Bigelow wondered whether there was a comparable phenomenon in human behavior; if there was, Wiener wrote, "Bigelow and I felt that we could safely go ahead with the treatment of the human links in the control chain as if they were pieces of feedback apparatus."

The two consulted Wiener's longtime friend, Dr. Arturo Rosenblueth, a Mexican neurophysiologist working at the Harvard Medical School. Rosenblueth informed them that there was indeed a pathological condition that produced similar actions in humans. It was called intention tremor, a condition often associated with brain injury or brain disorder. Rosenblueth's answer seemed to confirm the hypothesis that the human system, like a servomechanism, functioned by negative feedback. It was a small insight, but one with far-reaching consequences.

CROSSING CONCEPTUAL BOUNDARIES
Wiener and Bigelow presented their findings, and other researchers were then assigned to carry the work further. Wiener was happy to turn to the problem he was ultimately more interested in: the idea of applying the concept of feedback, heretofore considered only in terms of servomechanisms and engineering, to physiological systems — and, conversely, of applying physiological concepts such as "information" and "message" to the signals employed in engineering and in computing machines. In 1948, Wiener published a landmark book that made feedback the core of his theories about both human and machine behavior. He called the book *Cybernetics* (from the Greek word for "steersman"), thereby establishing a new, albeit amorphous, field of science. "It combines under one heading the study of what in a human context is sometimes loosely described as thinking and in engineering is known as control and communication," he wrote. To Wiener's surprise, *Cybernetics* was a best-selling science book. It served as a kind of introduction to computers and the prospect of increasingly "intelligent" machines — machines that one day might be able to communicate and learn as humans do. It also spawned a generation of researchers keen to investigate all aspects of the human-machine relationship.

By the mid-1960s, these advocates were establishing laboratories dedicated to the study of artificial intelligence, or AI, as the field of machine intelligence is known. In various parts of the country, researchers began to tackle — with varying degrees of success — the problems of duplicating the human intellectual skills of logic, reasoning and learning. In this "top-down" approach, early efforts focused on programs that could play checkers, later efforts on programs that attempted to translate natural human language, still later efforts on so-called expert systems — programs that include a fund of knowledge gleaned from a human expert and a logic-based system for making inferences. Today, roboticists are looking to build expert systems into space station robots *(pages 28-29)* and autonomous underwater vehicles *(Chapter 4)*.

The early days of AI saw the emergence of another research approach, called the "bottom-up" school. These scientists attacked the problem by trying to build neuron-like networks of electronic circuitry and then attempting to reproduce such human skills as vision and speech. In 1958, Frank Rosenblatt, a research psychologist at the Cornell Aeronautical Laboratory, simulated what he called a perceptron, an electronic model of the human brain, on an IBM 704 computer. In about a year, he predicted confidently, he would have a working model that would be able to recognize optical patterns. Moreover, Rosenblatt announced, perceptrons would eventually read and listen well enough to perform such tasks as translating languages and transcribing dictation. When a working model finally made its appearance in June 1960, it showed no such promise, in spite of an impressive array of photoelectric cells and yards of messy-looking wiring. "A laboratory curiosity" was one observer's dismissive assessment.

GOALS ON THE FAR HORIZON

Despite the overselling of perceptrons, bottom-up research continued, and many of today's scientists insist that it offers the best hope for achieving true machine intelligence. For them as well as their top-down counterparts, the pursuit of that dream has been greatly aided by electronic miniaturization. Computers can be fitted into the body of, say, a guided missile, a deep-space probe or a driverless truck. Yet robotic skills of sensing and thinking are rudimentary at best. In strictly controlled factory environments, for instance, robotic vision systems can execute simple inspection tasks such as verifying whether bottles are filled to the proper level. Still tantalizingly out of reach are programs that would enable a machine to make sense of the image in a useful way — to identify features and recognize objects under less than strictly controlled circumstances.

The quest for such skills is in no danger of abating, however. Sensory abilities such as vision and touch, while requiring expensive technology and highly sophisticated programming, would be extremely useful additions to a robot's feedback control loop. For his part, Norbert Wiener was far from sanguine about the potential uses and abuses of machines with ever-greater capabilities. Unless carefully managed, he warned, the second industrial revolution, as automation was touted, would likely "devalue the human brain, at least in its simple and more routine decisions." But Wiener also saw that much of the labor humans perform — moving parts from one place to another, inserting one part into another — is work much better done by machine. In factories around the world, robots incorporating the feedback control principles he and others investigated during World War II are relieving human workers of just those tasks — one answer to the centuries-old wish for human surrogates.

Giving Robots Sight

At its most sophisticated, machine vision might be described as a cross between artificial intelligence and automation. Not only must the machine capture an image, it must also analyze it to identify important features, interpret the results and decide on subsequent action based on the interpretation. Yet even at more rudimentary levels, vision systems enable robots to perform such tasks as installing automobile windows or making sure packages are lined up properly for labeling. They can also take over jobs that are too painfully detailed for humans to carry out with the necessary consistency, such as inspecting the hundreds of microscopic connections on semiconductor chips.

Robotic vision can be tailored for specific tasks. Whereas humans extract the meaning of the world around them from a whole battery of visual characteristics — color, texture and three-dimensional shape, for example — robotic systems can often operate usefully with only two-dimensional outlines in black and white.

With the exception of experimental optical processing systems *(pages 28-29)*, machine vision is a digital affair, recording a scene as varying intensities of light, each with a numerical value. To identify an object in some systems, the computer must compare the intensity at each point of the received image with corresponding points in a kind of template stored in memory — a task that requires enormous processing power even when the elements to be compared are kept to the absolute minimum. Other systems analyze the incoming numbers another way, comparing them with stored equations that reveal the relationships among various parts of the whole object *(pages 24-25)*.

Machine-vision systems work best in highly structured environments, such as assembly and production lines specially designed to accommodate them. In these situations, the pieces to be inspected or handled often appear at a prescribed rate — and sometimes in a prescribed orientation — and the lighting is controlled to avoid confusing shadows and to provide a constant level of light. As illustrated on the following pages, the more disorderly — that is, the more unstructured and unpredictable — the environment, the more sophisticated a robot's vision system is required to be. For some applications in the distant future, it will have to approach the daunting complexity of human vision.

A Simple Beam System

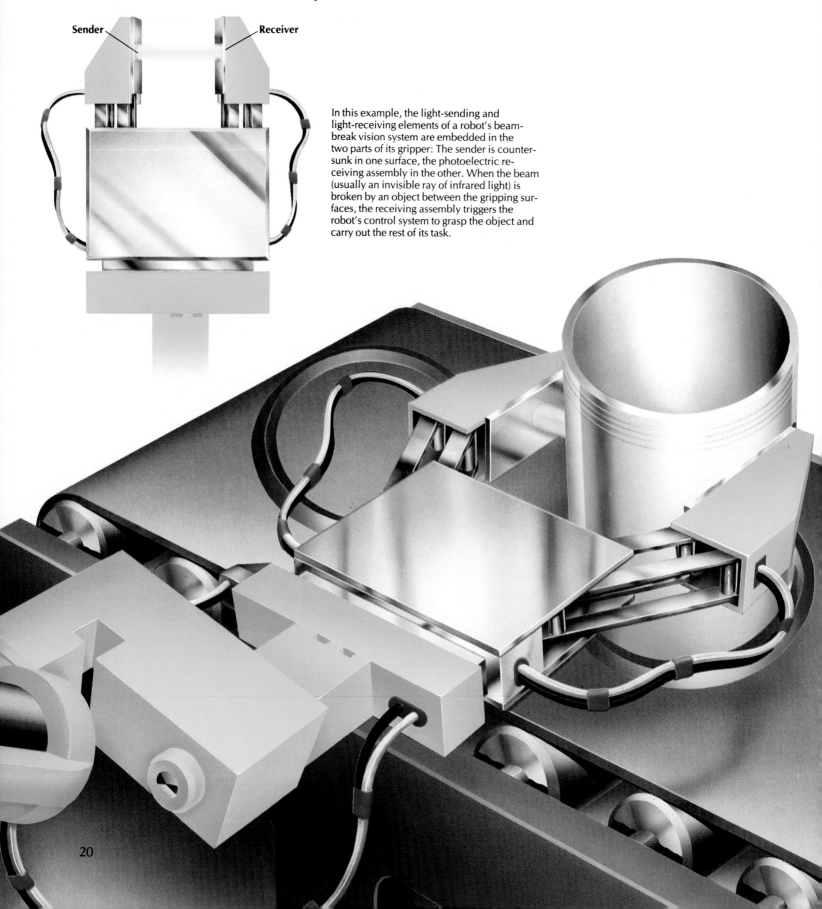

Sender **Receiver**

In this example, the light-sending and light-receiving elements of a robot's beam-break vision system are embedded in the two parts of its gripper: The sender is counter-sunk in one surface, the photoelectric receiving assembly in the other. When the beam (usually an invisible ray of infrared light) is broken by an object between the gripping surfaces, the receiving assembly triggers the robot's control system to grasp the object and carry out the rest of its task.

Among the simplest forms of machine vision are systems that employ photoelectric switches or sensors. One type, the so-called beam-break system, is a familiar feature of conveyor belts at supermarket check-out counters, which are designed to stop automatically when a product reaches the cashier. The system uses a light source, usually a light-emitting diode, to beam pulses of infrared light at a receiver — typically an assemblage that includes a photoelectric cell and various lenses, filters and amplifiers; the receiver responds by generating an electrical signal. In the case of the conveyor belt, as long as the system receives this signal, the belt continues to roll and groceries move forward; when an object passing between the light source and the receiver breaks the beam, the receiver assembly ceases to generate its signal, causing the belt to stop. (In some systems, the belt stops automatically if nothing breaks the beam for a prescribed length of time.)

The configuration and the hardware chosen for a beam-break system may vary with the application. In some cases, for example, the maximum distance between the light source and the receiver — known as the sensing distance — may be as little as a few millimeters; in others, it is as much as 30 meters. Special sensors also can be designed to work with transparent materials.

A beam-break system can provide a robot with useful — albeit rudimentary — vision, provided the robot's task is sufficiently uncomplicated and its environment sufficiently structured. For example, on an automated assembly line, such as the simplified one shown here, a robot may be assigned to handle one type of machine part and to perform one operation on it, such as picking it up and placing it somewhere else. The robot has no need to discern such details as the shape, texture or color of each part, because all parts are the same. The robot merely has to detect whether or not a part has appeared within reach of its grippers.

The environment in which this robot operates is kept as simple and orderly as possible: Identical heavy machine parts, identically aligned, appear in front of the arm. The robot is never confronted with any object except those it is supposed to pick up and transfer to another station for assembly. However, the robot might be programmed to signal a human operator if a gap appeared in the line.

An Eye for Flaws

In most cases, a robot designed to perform the task of quality control requires substantially more sophisticated machine-vision programming and hardware than it would need merely to detect the presence or absence of an object. Sometimes the robot system must also, in effect, be trained for the job, a process that involves showing the system varying examples of the desired object and letting the computer extract statistically significant properties. To inspect the gears illustrated here, for example, the robot first must locate and measure such characteristics as a gear's large central hole and the teeth around the circumference. As in all digital vision systems, such measure-

Having found a damaged gear with missing teeth, the robotic quality-control system pushes it into the reject bin. The next gear on the line *(center)* has been scanned by the camera; since it has no unacceptable flaws, it will be allowed to proceed. The gear under the camera's eye will be rejected for having only two small holes instead of three.

ments derive from variations in the intensity of the light reflected from the object and its background. This particular system permits no ambiguity in the environment: The gear must be set against a contrasting background, lighting must be arranged to avoid shadows and the gears must not overlap.

The light reflected from each gear is picked up by a camera and converted to a binary, or black-and-white, image. Using the geometry-based rules of so-called blob analysis, the computer separates object from background and locates other bloblike areas of dark or light, such as holes. The machine then takes a series of measurements of each blob. By comparing these measurements with the values extracted during the training period, the robot can reject a gear with broken teeth or one that is bent out of round beyond a given tolerance.

This vision system matches two sets of numerical data — one representing acceptable measurements, the other based on readings from the object under inspection. Here, the data includes the ratio of the gear's perimeter to that of the smallest ellipse that can contain it; the size, number and orientation of the three small holes; and the location of the large hole's center, which should be congruent with the center of the gear itself.

Small Hole

Center

Perimeter

Circumscribing Ellipse

Telltale Contours

In the real world, a robot must be able to deal with problems that arise from three-dimensionality. When one object is obscured by another, as it would be in a jumbled parts bin, for example, the robot must be able to identify the desired part and compute its orientation in order to pick it out of the bin. Human beings meet this challenge easily when they do something as ordinary as rummaging in a toolbox to find a hammer; the searcher needs only a few recognizable characteristics — a glimpse of the handle and one of the prongs on the head, say — to hypothesize where the rest of the handle might be.

Machine-vision systems capable of bin picking, as it is known, are so complex and slow that some researchers argue it would make more sense simply to reorganize assembly lines to eliminate disorderly bins. The counterargument is that this work method is too entrenched in factory routine to be done away with; moreover, it is exactly the kind of monotonous job robots are best suited to do.

The experimental system depicted here tackles the problem with a sensor that gathers depth data and a program that recognizes objects by locating clusters of three-dimensional features, such as corners, holes and cylinders. The sensor builds a contour map by scanning a plane of light across the scene and using triangulation to compute the locations of points along the intersection of the plane and the objects. This map, since it directly encodes the geometry of the scene, greatly simplifies the process of recognizing a desired object. Even relatively simple objects may have 20 or more distinguishing traits, but the feature-extraction program can function with as few as three: Upon finding one feature, the computer uses that trait's position as a guide to the location of a second feature; the two features are then used to predict the location of a third, thus fixing both the object's position and its orientation.

With an arrangement of a laser and a mirror, a bin-picking robot sweeps a series of fan-shaped planes of light across a bin full of parts *(below, right)*. A camera records the irregular line of light drawn by each plane, and the system computes the locations of points in the scene. The program looks for unique patterns of features — such as the characteristic edges and surfaces of a block *(diagram, right)* — to identify and locate the desired object. Then the computer positions and closes the gripper to pick it up *(left)*.

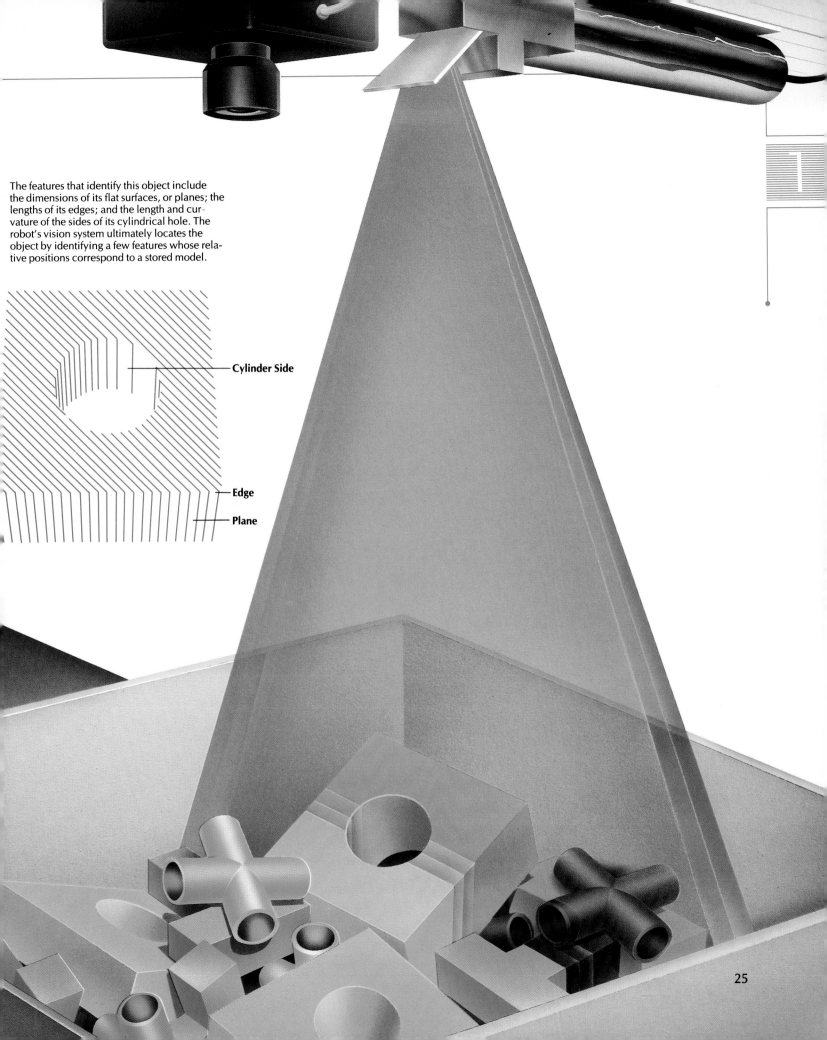

The features that identify this object include the dimensions of its flat surfaces, or planes; the lengths of its edges; and the length and curvature of the sides of its cylindrical hole. The robot's vision system ultimately locates the object by identifying a few features whose relative positions correspond to a stored model.

Cylinder Side

Edge

Plane

Digital Reference Scene

Camera

1 Using a zoom lens, the Tomahawk missile's DSMAC camera collects an image for comparison with a black-and-white digitized reference scene stored in memory. The DSMAC system allows the missile to check its actual position against its programmed one and to make any last-minute course corrections.

Navigating with a Matched Image

Outside the confines of an assembly line, a robot's need for adaptability increases. But vision systems that let a machine adjust its behavior to meet a changing environment are among the most complex and difficult to create. A Tomahawk cruise missile, for example, although not ordinarily thought of as a robot, employs two systems — one radar-based, the other optical — to guide itself to the target. The Tomahawk is designed to be launched from a surface ship, a submarine or a ground unit that may be several hundred miles from the missile's target. En route, the missile may sometimes fly as low as 100 feet above the ground to avoid detection by enemy radar. In so doing, it must also be able to avoid crashing into hillsides and to correct for so-called inertial drift, which could put it off course.

For most of its journey, the missile uses a TERCOM (terrain contour-matching) system to keep it on course. Stored in the missile computer's memory is a series of digital altitude profiles of strips of the landscape at certain places along the Tomahawk's intended flight path. The missile bounces radar signals off the landscape below and compares the real-time altitude profiles with the stored ones, making course corrections accordingly.

Later, when the missile is theoretically within a certain distance of its target, a DSMAC (digital scene-matching area correlation) system takes over from TERCOM. As illustrated here, the missile is programmed to survey the terrain at a given moment and to compare the resulting image with a digital reference scene made from a satellite reconnaissance photo — usually a distinguishing landscape feature, such as the section of riverbank shown at left. The missile performs a final realignment before heading for the target.

2 Having realigned itself after passing over the reference target, the missile may climb to a higher altitude so as to be able to drop directly on the target it is meant to hit — here, a railroad bridge.

27

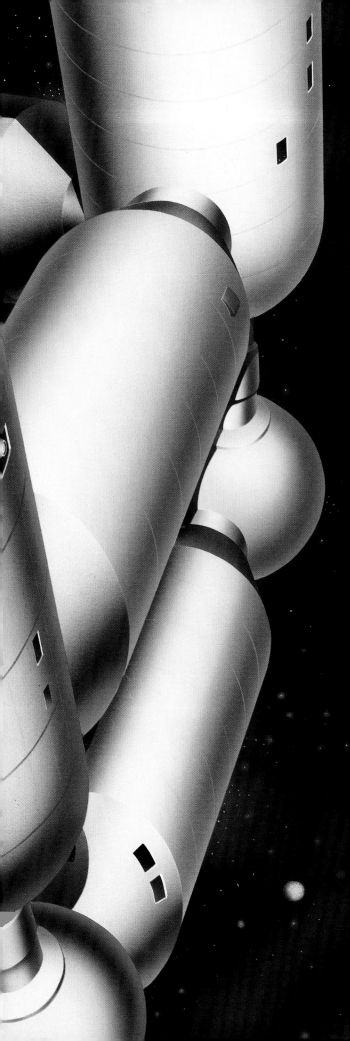

A Space Specialist

Future space stations in orbit around the earth are likely to include provisions for maintenance and repair robots responsible for routine upkeep. Initially, these handyworkers will be remote-controlled teleoperators *(pages 115-121)*, but the ultimate goal is to make them autonomous. Some researchers investigating vision systems for such robots have been attracted to a technology called optical processing, a form of computation that works by directly manipulating the analog information in light, rather than converting it to digital data first. If research-and-development efforts with optical processing are successful, space station robots might one day compute optical information fast enough to perform tasks similar to the minor repair illustrated here.

In this futuristic scenario, the robot has been instructed to replace an empty pressurized-gas canister in an exterior compartment. To do this, it must make its way to the site, remove the canister, put in a fresh one and carry the empty canister into the station for storage. In the course of this work, the robot continuously receives light patterns through its optical sensor for comparison with reference images stored in memory. These images are written onto a spatial light modulator — in effect, a kind of holographic transparency, or mask; each mask allows for multiple views of every object the robot is likely to see, from small pieces of station equipment to the earth, looming hundreds of miles away.

An arrangement of lenses conjoins the mask and the light from the external scene. At any point where a stored image appears in the current scene, the system registers a bright peak of light. Guided by very sophisticated software known as a librarian program — which quickly narrows the robot's search for the best possible match — the vision system could check its input against as many as 1,000 masks per second.

A space station robot of the future uses one of its three arms to grip the space station while the other two replace an empty pressurized-gas canister. In order to navigate its way to the compartment, the robot had to use its optical processing system — retrieving and comparing thousands of images in a few seconds — to recognize such objects as the earth, nearby spacecraft, other robots, human astronauts and portions of the space station itself.

Scouting New Territory

Cruise missiles and robot handyworkers on a future space station are intended to perform well-defined tasks in a particular and, to some extent, predictable environment. Although the systems needed to guide them are complex, such robots can be provided a finite and manageable store of reference images against which to compare reality. But the creation of a robot whose job is to navigate in unexplored and potentially dangerous territory is a challenge of immense proportions. For example, a military autonomous land vehicle (ALV) designed to scout behind enemy lines, clear a mine field or serve as a forward reconnaissance post would need a fund

To navigate in this rural landscape, an autonomous machine would need a vision system capable of detecting and identifying objects in real time — fast enough so that the vehicle does not drive itself off the road or into a tree. In addition, it would have to be supplied with a store of information that would seem self-evident to a human: for instance, that those fuzzy, four-legged objects might wander into the road but the large, hard-edged geometrical one will remain stationary.

Robotic vision systems often begin to study a scene by searching for the edges of objects. By locating abrupt changes in light intensity, the computer creates an edge map of objects that stand out from their backgrounds as these trees stand out against the grass and sky.

of initial knowledge that would allow it to recognize, first, that a green, leafy object in the afternoon sun and a dark gray silhouette in the twilight are both trees, and second, that they are the same tree.

The human ability to identify objects in a cluttered scene by combinations of shape, color, texture, depth and other visible traits can, to some extent, be replicated in machines, although no present-day vision system even approaches human prowess. A robot that can extract the meaning of an unfamiliar scene as quickly as a person will require prodigious computing power, plus some version of the common sense and recol-

lection that humans bring to bear on everyday perception.

Shown here is one extremely futuristic approach to the problem of real-world vision for an autonomous vehicle. It employs a system of multiple TV cameras, a laser range finder, sonar sensors and other instruments that combine their input on a so-called computer blackboard to build up a composite image of the robot's environment. In addition to this purely visual component, the vehicle would also need learning programs and expert systems — creations from the domain of artificial intelligence — to interpret and act upon incoming data.

To find how far away and how big the barn is, the vehicle needs depth perception. One approach employs two cameras to record the scene from slightly different viewpoints, then converts the images to a pair of digital edge maps. Discrepancies between the two maps are plugged into formulas to calculate depth. The vehicle could also use sonar or laser range finders for coarser but faster depth perception.

To locate the edge of the road in order not to drive off it, the robot might use the technique of texture analysis. Textures are represented in machine-vision systems as regular patterns of brightness values. By noting sudden changes in these patterns, the system can detect a change in the physical surface — from the gravel road to the near edge of the grassy field to the field itself.

Often the easiest way to distinguish one object from another is by color — white sheep against a green field, for example. An autonomous vehicle might perceive color by using a camera to record the red, green and blue components of the light reflected from each point of the scene. After adjustments for ambient light, a deep green might register as having no red and very little blue; white, in contrast, would contain equal parts of all three colors.

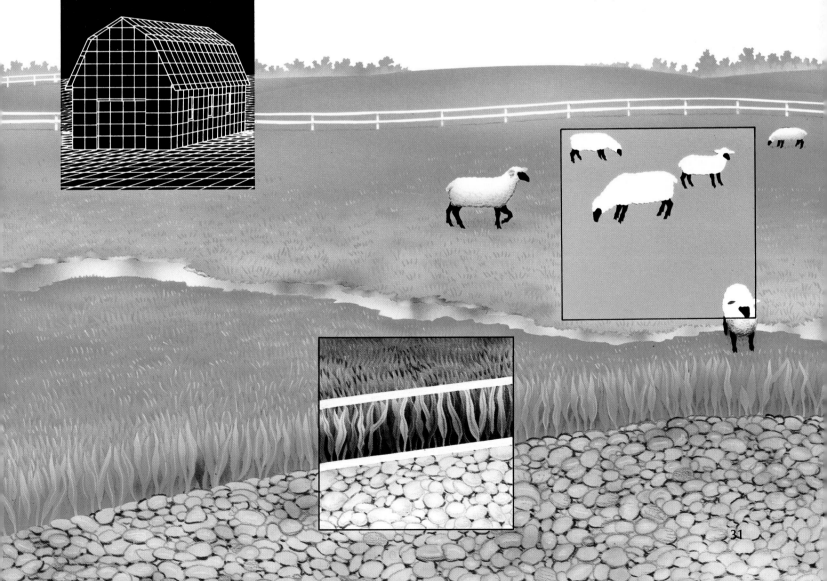

31

A Steel-Collar Work Force

Four miles from the foot of sacred Mount Fuji, the bright yellow walls of the Fanuc factory rise from a pine forest, incongruously modern in a landscape that symbolizes traditional Japan. But this manufacturing facility is even more modern than externals suggest. Fanuc makes robots, and it does so with a remarkable work force — 110 human workers laboring alongside 300 robots. The robots and humans work together for eight hours on one shift; then the humans go home and the robots work another 16 hours with a single human supervisor.

The robot-making robots of Fanuc and their fellow "steel-collar" workers in factories around the world bear no resemblance to the walking, talking androids of science fiction. Most are blind, deaf and mute. Leaning over their tasks, they paint, weld or assemble parts for as long as necessary to get the job done. Taking no time off for sleep or coffee breaks, they can toil around the clock, oblivious to noise or poisonous fumes. Despite their limitations, robots are leading the world into an era of great economic and social change.

The course of the changes can only be guessed at. Already the dream of the totally automated, computer-run factory is approaching reality. In many industries, robots can make goods cheaper and better than is possible by traditional means. Inevitably, they have usurped the jobs of some human workers in factories, and that process shows every sign of accelerating. The effect on society will be complex. In the opinion of James S. Albus of the National Institute of Standards and Technology (formerly the National Bureau of Standards), the world is "poised on the brink of a new industrial revolution that will at least equal, if not far exceed, the first Industrial Revolution in its impact on mankind."

Robots came to the factory first in the United States, thanks largely to the vision of a tireless entrepreneur named George C. Devol Jr. and to the enthusiasm and tenacity of an engineer-supersalesman named Joseph F. Engelberger. The pair developed pioneering machines, worked to create a market for them and established a company that became one of the country's largest suppliers of robots. However, it is not in the United States but in Japan that the robotic revolution has progressed the furthest.

In Nagoya, 150 miles southwest of Tokyo, Brother Industries employs 25 robots to build electronic typewriters in scores of varieties, altering designs as needed to accommodate the different keyboard requirements of the world's languages, professions and national preferences. Robot assembly is so efficient and flexible that Brother can profitably manufacture these many different sorts of typewriters in batches as small as 10.

The Fanuc plant in the shadow of Mount Fuji can add 300 more robots to the world's force of steel-collar workers every month. At that manufacturing facility, unmanned carts are programmed to carry rough metal castings from an automated warehouse to a production area. There, robots unload the raw materials and feed them to computer-controlled machine tools, which cut and shape parts automatically, according to numerical code. Then finished parts are transferred back into the carts for a trip to the warehouse, where they are kept until needed by the robots on the assembly line.

Elsewhere in Japan, Mitsubishi once developed a robot to sort a fishing boat's catch, and Hitachi has made one that can assemble vacuum cleaners. By the end of the 1980s, some 175,000 robots were at work in Japan, about two thirds of the world total. The rest of the world has been rushing to catch up:

• In Atlanta and Chicago, lines of Kuka 662-100 robots complete 98 percent of the spot-welding at two Ford Motor Company plants.

• Outside Turin, Italy, Digital Electronic Automation put six Faber C-5000 robots to work under the supervision of two operators. The robots assemble 140 transmission shafts an hour, matching the output of 15 human workers.

• In Fort Worth, Texas, General Dynamics installed 11 Cincinnati Milacron T^3 robots to help build F-16 fighter planes; the T^3 can select its own tools from a rack, drill holes accurate to .005 inch and machine the perimeter of 250 different parts.

• Near London, Egerton Hospital Equipment put a Dainichi-Sykes PT600 to work welding the frames of hospital beds.

• In Poughkeepsie, New York, IBM set up a robot to load and unload a computer that writes code onto blank disks. The robot knows which of four machines gets a specific blank disk, when to remove the recorded disk, how it should be labeled and which container to pack it in.

Today's industrial robots, so efficient at these varied tasks, are fundamentally different from machines of the past. They do more than imitate a human skill, as, for example, a sewing machine does. A traditional sewing machine makes neater stitches than a human being normally does—and sews them faster. But that is essentially all it can do. Most industrial robots are multifunctional and reprogrammable. A robot may be programmed in advance to perform similar tasks with varying requirements (such as drilling several holes of different sizes) or to change from one task to another (such as drilling a hole, then placing a bolt into the hole).

The antecedents of these remarkable machines can be traced back at least to classical Greece, when Hero of Alexandria built simple devices to demonstrate the principles of hydraulics. But the key elements—programmability and multiple skills—did not appear in rudimentary form until the 19th century. In 1804, French inventor Joseph Marie Jacquard unveiled a loom that altered the weaving industry and foreshadowed a key capability of computers. It was programmable, using punched cards to automatically control the weaving of complex patterns; changing a card changed the design produced. A generation later, in 1830, the American Christopher Spencer built a lathe that could be pro-

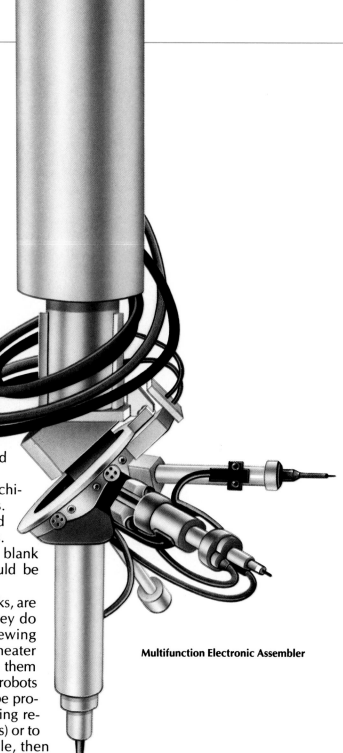

Multifunction Electronic Assembler

grammed to perform a few different kinds of work. Its movements were controlled by cams; depending on which cams were set into the control unit, the lathe would make screws, nuts or gears. But another 124 years would pass before George Devol's patent for Program Controlled Article Transfer — an automated system for shifting components from one place in a factory to another and for manipulating them in various manufacturing processes — ushered in the age of the industrial robot.

INVENTOR ON THE MOVE

The bespectacled and voluble Devol is a maverick among modern engineers. A latter-day Edison, he is a lone inventor, spinning off patentable ideas and moving from one field to another as opportunity arises. He has acquired more than 40 patents on inventions that range from automatic laundry machinery to electric doors. He manufactures few of these devices; the successful ones are produced by companies in which he has an interest.

Devol was born in 1912 to a well-to-do family in Louisville, Kentucky. Like Edison, he displayed his talent early: While still a teenager, he built and operated an electric generating plant for his school. He skipped college and, at the age of 20, went straight into his career as an inventor-entrepreneur. With money raised from friends and his banker father, he set out to make better sound-recording and playback equipment for the then-burgeoning talking movies. Bigger competitors quickly discouraged him from that goal, but one element in his proposed playback mechanism — a photoelectric cell — seemed useful for other purposes. He adapted it to develop one of the first automatic door openers — a device similar to those widely used today in supermarkets and elevators. That led him into automatic controls, and he patented an application of the photocell to control the placement of labels for printing and cutting.

During World War II, Devol formed and ran a company that

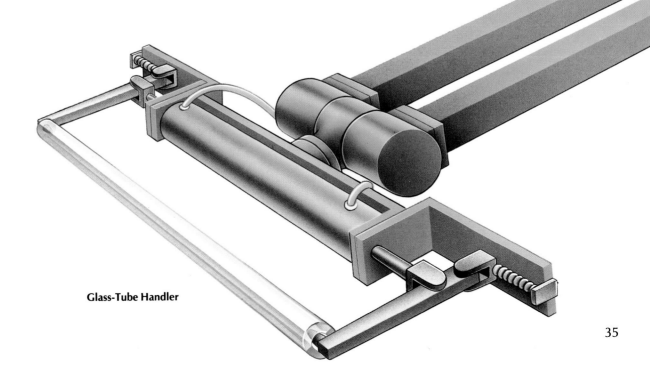

Glass-Tube Handler

manufactured radar-jamming equipment. ("On D-Day," he recalls, "the Germans had radar all over the Normandy beachheads and a lot of other places to detect ships as well as aircraft. The Allies used quite a lot of our equipment and were able to knock out the effective use of the radar.") But he soon left this business to pursue a new idea. He saw that a magnetic recording could make a unique control system for a machine. His idea seems simple in retrospect: Equip a machine tool, such as a lathe, with sensors that would relay to a magnetic recorder signals indicating each movement of the lathe cutter. Then have a skilled machinist use the lathe to make the desired product — say, a gear of a specific design. The human machinist would need to do this only once. Thereafter, the magnetic recording of his actions would guide the lathe to reproduce that same gear repeatedly. The lathe thus became programmable, and this method of programming — employing a human operator to teach the machine its job — continues to be popular for today's industrial robots.

Devol got several patents on the magnetic control but was unable to excite commercial interest at the time. His magnetically programmed lathe never was built, and Devol searched for a broader application of the notion.

He knew that, contrary to popular misconception, less than half of the world's goods were mass-produced. The rest were made in batches too small to justify special automatic machines for each step in their manufacture. Even mass-produced things, such as automobiles, required many human workers performing boring, unskilled tasks that seemed to belong more properly in the realm of a machine: The workers did nothing but putting and taking as they moved objects from place to place, feeding parts to machines, assembling them into products, packing the products in boxes. What industry needed were machines that could do one thing for a while, then another, then yet another. Such a machine might, for example, put English-character keys into a few typewriters, then put Cyrillic keys into another few and Arabic keys into a third batch.

CONTROL FOR AN ALL-PURPOSE MACHINE
In 1954 Devol filed for U.S. Patent No. 2,988,237 on a control system for a single, all-purpose machine that could be programmed as needed to complete a variety of jobs on the factory floor — Program Controlled Article Transfer, in his summary phrase. Devol and others would later build on this idea to create an industrial robot that showed a distinct kinship to a human arm although it was, of course, power-operated and automatically controlled (pages 47-59). It had a shoulder, upper arm, forearm and hand. The hand consisted of one or more replaceable tools, such as a saw, a drill, a screwdriver, a welding gun or a gripper for pick-and-place operations. Joints at shoulder, elbow and wrist hinged, swiveled or slid to get the hand into working position. The number and type of joints determined the robot's "degrees of freedom" — the different ways it could move. Many robots have six degrees of freedom. Three permit the machine to move up and down, back and forth, and side to side; another three permit the wrist itself to rotate, shift side to side, and move up and down. By comparison, a kitchen blender has one degree of freedom; it can spin in only one direction.

Mechanical arms, even power-operated ones, had been available before Devol came along. His uniquely valuable contribution was programmable automatic control. But when he tried to interest manufacturers in the machine, he was rebuffed. "You'd go around and tell someone you wanted to put a robot in their plant," he recalled later, "and they'd almost call the guards to come and get you." In 1956, he found a kindred spirit in Joseph Engelberger, then the 31-year-old chief engineer of the aerospace division of a Connecticut company, Manning, Maxwell and Moore. Engelberger listened intently as Devol argued that "50 percent of the people in factories are really putting and taking" and then went on to describe his plans for a programmable manipulator, a machine "that will put and take anything."

As an engineering student at Columbia in the 1940s — where he had taken the university's first course in servo theory — Engelberger had read the stories of Isaac Asimov and had been fascinated by robots ever since. Recognizing Devol's manipulator as a robot, Engelberger was caught up by the concept. "It seemed like a good idea in the cold, gray dawn. And, I guess, the thing that slowly struck me was that the technology for doing it was close at hand."

MISSIONARY FOR A ROBOT
Thereafter, Engelberger became perhaps the world's foremost proponent of robots. He persuaded his employers, Manning, Maxwell and Moore, to license Devol's patents, but after little more than a year the company's interest cooled, and it ordered Engelberger to fold his robotics section. He quit, and over the succeeding years he worked with Devol to persuade a succession of corporate backers to support their infant enterprise, which eventually became known as Unimation, Inc. Not until 1974 — 20 years after Devol filed for his patent — did Unimation earn a profit.

When Engelberger walked out of his job with Manning, Maxwell and Moore, a great many engineering problems remained to be solved before Devol's robot was truly ready to go to work. The stickiest related to accuracy. In order to be useful in a factory, the machine had to be able to position its gripper, sweeping at the end of a nine-foot arm, to within 1/1,000 of an inch of any desired point. This, in turn, depended on the control unit's knowing where the gripper was at all times. Human beings possess such a skill, provided by nerves that keep the brain continuously informed of the position of each part of the body. (You can, for example, make your finger touch the arm of your chair without looking.) Devol's robot would have to imitate the human ability by using internal sensors to detect how much each joint had moved from its previous position. Then the control unit could compare the new position with the desired one and move the joint ahead or back as necessary. Existing sensors were not reliable enough for use on the robot, but workable versions were eventually devised.

Moving the joint so precisely proved to be as difficult as monitoring the action. The force for the movements was hydraulic — delivered by a piston that was pushed by the pressure of oil, like hydraulic brakes on a car. Seals on the piston had to press very firmly against the wall of the surrounding cylinder to avoid leakage. But if they pressed too hard against the cylinder, the piston would not be able to move in tiny increments. After much experimentation, various types of nylon and Teflon® seals were found to be both leak-resistant and capable of smooth, accurate movement.

By 1959 Unimation's engineers, aided by Devol, had overcome enough of these obstacles to build a prototype, a heavy-duty machine about the size of

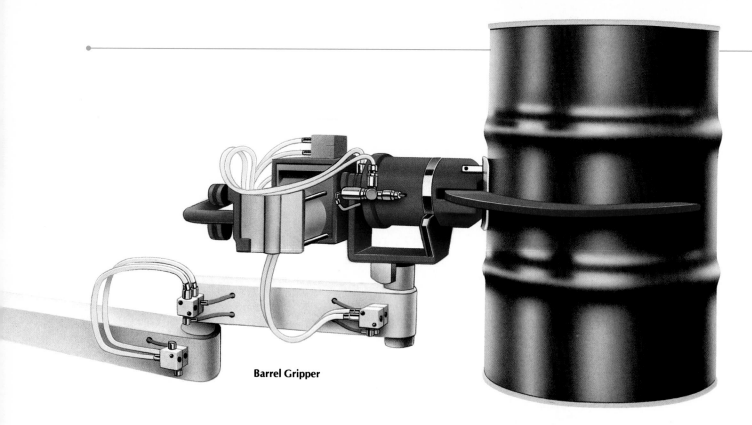

Barrel Gripper

a golf cart. General Motors bought it for a trial, installing it in a Trenton, New Jersey, plant to pluck hot parts out of a die-casting machine and quench them. Despite this sale, struggles still lay ahead for Unimation. The company received only about 30 orders over the next three years. Relatively low wage rates for American workers at that time provided little incentive for investing in expensive robots.

Engelberger embarked on an aggressive campaign to convert American industry to belief in the robot, which had been named Unimate. He was well suited to the role of missionary. Outgoing, a wide smile ever present, he affected items of dress that suggested a passion for efficiency: slip-on shoes and a clip-on bow tie. Of the tie he once said, "Clipping it on saves two minutes every morning."

Engelberger made the most of the image he cultivated. He wrote articles, gave speeches and exhibited the Unimate at trade shows. He also proved to be a master of publicity. At a press conference at New York's Biltmore Hotel, he arranged for a Unimate to fill in as a one-armed bartender. He became a regular on television talk shows, appearing 25 times over the years, often with a Unimate to perform tricks. On one late-night program, he recalled, "we washed windows, because the gag line is that you can't get a maid to do windows, right? So we did. We had the robot open up the curtain, wash the window, dry the window, unlock the window, pick up a water can, water the flowers outside, put it down, close the window, close the curtains."

Engelberger even spread the gospel abroad. In 1967 the Japanese government invited him to tour their country and to speak to a group of engineers in Tokyo. "The Japanese were predisposed to robots," he said, although even he was surprised by the enthusiastic welcome he received. Accustomed to speaking to groups of eight or 10 at home, he looked out from behind his lectern into the attentive faces of 700 Japanese engineers. Despite the impressive attendance, Engelberger's government hosts were concerned that no one would ask questions at the end of his lecture. To spare their guest an embarrassing silence, they had prepared a list of three or four items they might ask about. This precaution proved unnecessary. Engelberger's eager audience urged so many questions that

Suction-Cup Lifter

the session ran on for five hours. Within a year of Engelberger's brief tour, Unimation had licensed Kawasaki Heavy Industries to build Unimates in Japan.

By this time, Engelberger's campaign had begun to pay off, in the United States as well as abroad. In 1966 General Motors ordered 66 Unimates for welding automobile bodies at its Lordstown, Ohio, plant. Although that was the biggest sale Unimation would make for eight years, applications of robots steadily increased during the late 1960s and early 1970s, spurred both by spiraling wages and by developments in computers and microprocessors that greatly improved automatic controls. New types of robots appeared. Among the most successful was one that—unlike the 4,000-pound, long-armed, hydraulically powered Unimate—weighed only 120 pounds, had a reach equivalent to that of a human worker and operated its joints with electric motors controlled by computer.

If this smart midget robot was different from the Unimate, so was its inventor different from George Devol. He was Victor Scheinman, educated in the artificial intelligence laboratories at M.I.T. and Stanford University, who has shuttled back and forth between academia and private enterprise. Scheinman initially planned to create a tabletop robot not for industrial use but as an aid to research in artificial intelligence—a "unit that a student could have on the corner of his workbench," he later said. Its industrial applications became obvious, however, and he soon formed his own company and set out to market the machine, despite some naïveté about trade customs. One sales attempt attracted perhaps more attention than he had bargained for. Planning to exhibit the robot at the Fifth International Symposium of Industrial Robotics in Chicago in 1975, he simply packed his things and went. But he had neglected to reserve a display booth at the symposium, and as Engelberger described the scene, the young inventor found himself "standing outside with his nose pressed against the window."

Engelberger knew Scheinman, who was in a sense a Unimation protégé—he had attended Stanford on a fellowship sponsored by Devol and had visited the Unimation plant. So Engelberger invited Scheinman to share his own booth. "What the hell," he told Scheinman. "You're not going to hurt us." Other exhibitors, wary of added competition, were less than pleased by Engelberger's hospitality. Within an hour, Scheinman was again out of the exhibition hall. Not easily discouraged, he set up his robot on the front steps and proceeded to demonstrate it to any interested passersby. As it turned out, the whole adventure did lead to a sale; a year later Unimation bought up Scheinman's company and with it the little manipulator.

Investing in the small robot proved profitable for Unimation, because a big customer soon appeared. General Motors completed a survey of its operations and found that 35 percent of its workers were involved in subassemblies — putting together the building blocks of a vehicle, such as the dashboard or the window cranks. In addition, they discovered that about 90 percent of the parts in an automobile weighed less than five pounds. Plainly, there was plenty of work for robots that could deal with small components.

The General Motors engineers — some of whom may have witnessed Scheinman's alfresco performance in Chicago — asked for bids on what they named PUMA, a Programmable Universal Machine for Assembly. The specifications called for a robot no bigger than a human being, which could operate alongside human workers for light assembly jobs. The Scheinman arm, slightly modified, met the requirements, and by 1978 the first one was installed at GM's Technology Center in Warren, Michigan.

Other manufacturers soon followed GM, pressed by the sharply rising costs of human labor. By the early 1980s, human workers in the automobile industry were earning more than $20 per hour, while the cost of operating a robot had dropped, by some estimates, to less than $6 per hour. Inevitably, manufacturers began to turn to steel-collar workers. By mid-decade, robots were being produced by 230 firms worldwide, including some, such as General Motors, that were themselves major users of robots. Approximately 100,000 of the machines were laboring in work forces around the world. In one 12-month period the Japanese nearly quadrupled their supply of robots, and in the United States the number doubled between 1982 and 1984. By 1990, the annual rate of increase in the U.S. total was about 3,000 robots, according to one estimate, and by the end of the century, the nation's robot population may increase to more than 60,000 machines.

WHERE ROBOTS STAR

Many of the robots now in use do jobs that are especially onerous for human workers: those that require great strength, pose danger or are prone to simple but costly error. Warren P. Seering, a professor of mechanical engineering at M.I.T., cites spot welding as a typical example. Spot welding stitches together metal parts with a gun that releases a powerful spurt of electric current, fusing the metal pieces at each point it touches. The welding tool used by a human worker weighs 100 pounds or more, and although it is generally mounted on a counterbalancing frame, it is difficult to handle. Even for a husky human welder, the job is exhausting.

For spot-welding automobiles, said Seering, "the robot does not have to interact in a carefully controlled way with the piece of work. All it need do is anchor the spot-welding tool roughly to a part already lightly welded in place on the car and then apply pressure and electric current." He added, "Robots are the solution of choice in this case because they can do what humans cannot: produce and control large forces." Strength and tirelessness, of course, are useful in a number of jobs. As mechanical supermen, robots may be called upon to do anything from shifting heavy components between workstations on a factory floor to toting bags of cement.

Spray painting (pages 47-57), observes Professor Seering, is another task suited to robots because "robots do not need to breathe." Unlike human painters, they are unaffected by the poisonous fumes in the spray booth. "Robots are better at this task," Seering says, "not because they are faster or cheaper than humans, but because they work in a place where humans cannot."

Third in Seering's list of useful jobs for robots is the assembly of electronic parts. Robots shine at installing chips in printed circuit boards "because of a capability that robots have and people don't: A robot, once properly programmed, will not put a chip in the wrong socket." This automatic accuracy is particularly valuable with circuit boards, which generally cannot be tested until all the components are in place. If the wrong chip has been inserted or one has been put in backward, the board will not work. Locating and fixing such mistakes is costly. Thus, Seering said, "robots have the advantage in 'board stuffing' because the boards they produce have a much greater likelihood of working."

A SHOOT-OUT BETWEEN PAINTERS

Even in areas of clear-cut superiority, however, robots have sometimes performed erratically. When General Motors' Hamtramck, Michigan, assembly plant opened its doors in 1985 it was supposed to be a showcase of high technology, where an army of 260 robots would help the 5,000 human workers turn out 60 new cars per hour. One year later the factory was still producing only 35 cars an hour, largely because of malfunctioning of its automated spray-painting system. At times the computer-controlled paint booths became the scene of a high-tech shoot-out as robot painters drew a bead on one another instead of on the cars. Several hundred cars were so badly painted that they had to be shipped to a nearby Cadillac plant — a factory built in prerobotic 1929 — to be repainted by human workers using old-fashioned spray guns.

A Ford assembly plant in St. Louis had to fight a different and even more vexing problem. Engineers had great difficulty getting the factory's computer-controlled machines, produced by two dozen different suppliers, to communicate with one another. Ford's equipment was so complicated that human workers could not operate it effectively, even after months of training. The glitches caused a seven-month delay in the introduction of a new line of vans.

At Buick City in Flint, Michigan — part of GM's $40-billion investment in automation — robots were put to work installing windshields. Unfortunately for Buick, the robots lacked proper depth perception, and they shattered many of the units.

In some cases, the problems were not simply the bugs to be expected when any new kind of machine is installed: The robots proved unable to compete with human workers. Professor Seering described a test of one of General Motors' PUMAs at the seemingly elementary task of inserting light bulbs in automobile dashboards. The robot was programmed to pick the bulbs one at a time from precisely arranged racks, then swing to a socket and push and twist the bulb into place. Because the robot was not capable of sensing that the bulb was sliding into the socket, it had to work blind, moving at a very slow pace to follow its programmed path precisely. By contrast, a human worker simply grabs a handful of bulbs from a bin and quickly twists one after another into sockets with finger and thumb, easily beating the robot.

Such limitations are being overcome in newer types of robots that began to emerge from research laboratories in the 1980s. Unlike earlier machines, they are not blind and deaf but do their job with the aid of sensors. Many are fitted with video cameras or other visual detectors. For example, Japan's Nissan Motor Company has begun inspecting the paint finish on cars with robots that are equipped with a laser and special optical scanners. The robots can scan the entire painted surface of a car in little more than a minute and spot irregularities as small as 1/100 of an inch across. At the same time, the robots note the size, location and number of surface irregularities, and they can measure the gloss of the finish.

Even robots engaged in simple pick-and-place work can sometimes do their job better with sensory assistance. For example, a chocolate-packing machine devised by Unimation employed a visual sense. The job was sheer drudgery for human packers. "Women sit at a table," explained Joseph Engelberger, "and a line of empty chocolate boxes moves along in front of them. Canned chocolates come toward them on a white conveyor belt. The women put two chocolates per second into the box. That's their job for eight hours a day." He added, "The manufacturer says no one has lasted two years on that job. They go bonkers." Engelberger's solution was two small arms, their fingers fitted with silicone suction cups that could lift the candy "as gently as the women with the white gloves did." A video camera told the arms where to find the chocolates: The vision system had to do no more than recognize black objects on a white background.

Even as the capability of individual robots has grown, the power of the computer has harnessed them more effectively for group action. A technology called computer-integrated manufacturing, or CIM (pages 44-45), orchestrates virtually all stages of work in a plant, including the operations of robots. CIM generally begins with computer-aided design (CAD), in which an engineer uses a computer first to generate a blueprint for a new product or component, then to modify the design and even test its real-world performance against computer-simulated stresses. CAD technology allows engineers to debug a design — or improve an existing one — without building prototypes. Engineers at the Fisher Body Division of General Motors, for example, have used CAD techniques to evaluate the performance of car-body designs as they are subjected to the twists and bounces of a computer-simulated roadway. All specifications can then be

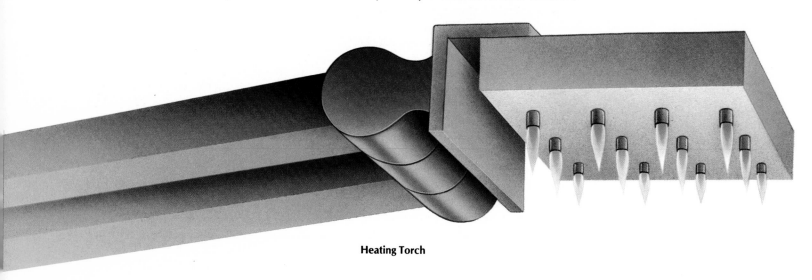

Heating Torch

42

Ladle

stored in the computer's memory for future use. On command, such computer-stored specifications can be tapped for computer-integrated manufacturing, which gives managers and technicians close control over every phase of production.

The computerized control system running a CIM facility is a decision-making hierarchy, typically with five levels. Described anthropomorphically, these automated levels are: the factory coordinator, which issues a plan dictating how many of which product to make that day; shop coordinators, each controlling production in a given department; line supervisors, which contain libraries of programs for assembling different versions of the product; workstation operators on the production line, which regulate machines that carry out a particular task; and equipment operators, the robots that perform the actual work.

The benefits of CIM all derive from the computerized management of information. Often, less time is needed to make the product. Inventory is minimized. And, because CIM can place a factory's entire output at the call of one master computer, products can be made efficiently in small batches. Allen-Bradley Company, Inc., a manufacturer of industrial controls based in Milwaukee, operates a computerized factory that can turn out different versions of a product in lots as small as one unit — and make them at mass-production speeds and costs.

The Allen-Bradley system is set up to begin each day at 5 a.m., when the company's main computer pulls from its memory the customer orders received the previous day and relays them to a master scheduling computer. At 7:30 a.m., the scheduling computer turns on the assembly line and tells it what to make. Conveyors and robots begin to move with a rush of pneumatic power, whooshing, whistling and flashing their lights. Plastic casings the size of pocket radios begin to roll through 26 assembly stations. At the first, a robot pastes on each casing a computer-printed label, a machine-readable bar code like the ones on packages in the supermarket. At each succeeding assembly station, the labels are read to tell robots which parts to put into which casings, ensuring that the assembly proceeds exactly as called for by the computer order. Finally, a laser printer zaps product information onto the casings, which are then automatically packaged, sorted by customer order and shunted off to the shipping dock. This automated line can produce 600 units per hour — not all alike but in any mixture of two sizes with 1,000 possible combinations of parts. It can, if necessary, make 599 of one model, then one of another.

The Allen-Bradley line requires just four human technicians, mainly standing by to clear jams. Obviously, robots displace people, taking over jobs that once required human beings. And the impact on the labor force, already considerable, will grow in the future as more versatile robots become available. Experts differ, however, as to the ultimate severity of this impact. According to one fairly dire estimate by Robert Ayres and Steve Miller of Carnegie Mellon in Pittsburgh, millions of workers will lose their jobs to robots in the United States alone. They

Managing an Automated Factory

With computer-integrated manufacturing (CIM), individual robots are meshed in a factory-wide system, and virtually all work is orchestrated by a network of linked computers. Typically, CIM is a three-stage process, beginning at graphics terminals where humans and computers collaborate on the design of a new product. Other computers then plan every step of production, down to the smallest detail. Finally, the plans are carried out by robots and machine tools on the factory floor.

The manufacturing process is usually controlled by a five-level computer hierarchy. Computers at each level are programmed to do certain jobs and make certain kinds of decisions. As in human hierarchies, orders flow from top to bottom, but vital information can move in either direction.

At the summit of this electronic pyramid is a level that can be described as the plant coordinator, which compiles and hands down a daily production plan. Broken into directives for various factory departments, the plan is digested and simplified as it passes through successive supervisory levels. The equipment operators — the robots at the base of the pyramid — follow commands issued from above to perform the actual labor of the factory, milling parts or, as illustrated here, installing electronic components. In some plants, two or more levels of control may be combined in a single computer.

By managing the flow of information efficiently, CIM can significantly reduce production time. A company can respond quickly to the demands of the market, without having to maintain large and expensive inventories of a product. Special orders, even for a single item, can be filled speedily and profitably because the factory's robots can switch from one task to another in an instant. The constant flow of information between levels also enhances quality control. Since monitoring is continuous, with tests and inspections at every step, a defective article is usually discovered before it leaves the plant.

The Control Hierarchy

The Plant Coordinator is responsible for overall planning. While translating orders from distributors into daily production goals, it also directs progress toward long-term objectives. It keeps track of parts and product inventories as well as handling orders and billing.

The Shop Coordinator schedules production and allocates resources within a given department. It also collects production data, compiles a daily production report and stores customer shipping information.

The Line Supervisor, an automated foreman, oversees the activities on a given line. It passes instructions to each workstation and reschedules production whenever a piece of equipment malfunctions.

The Workstation Operator controls the actions of one or more pieces of automated machinery programmed to perform a specific task. It also oversees routine maintenance duties, such as the replacement of worn or broken tools.

The Equipment Operator, at the bottom of the CIM hierarchy, does the nuts-and-bolts, shop-floor work of manufacturing. It places components, tightens screws and moves heavy loads while its sensors monitor work in progress to make sure the job is done correctly.

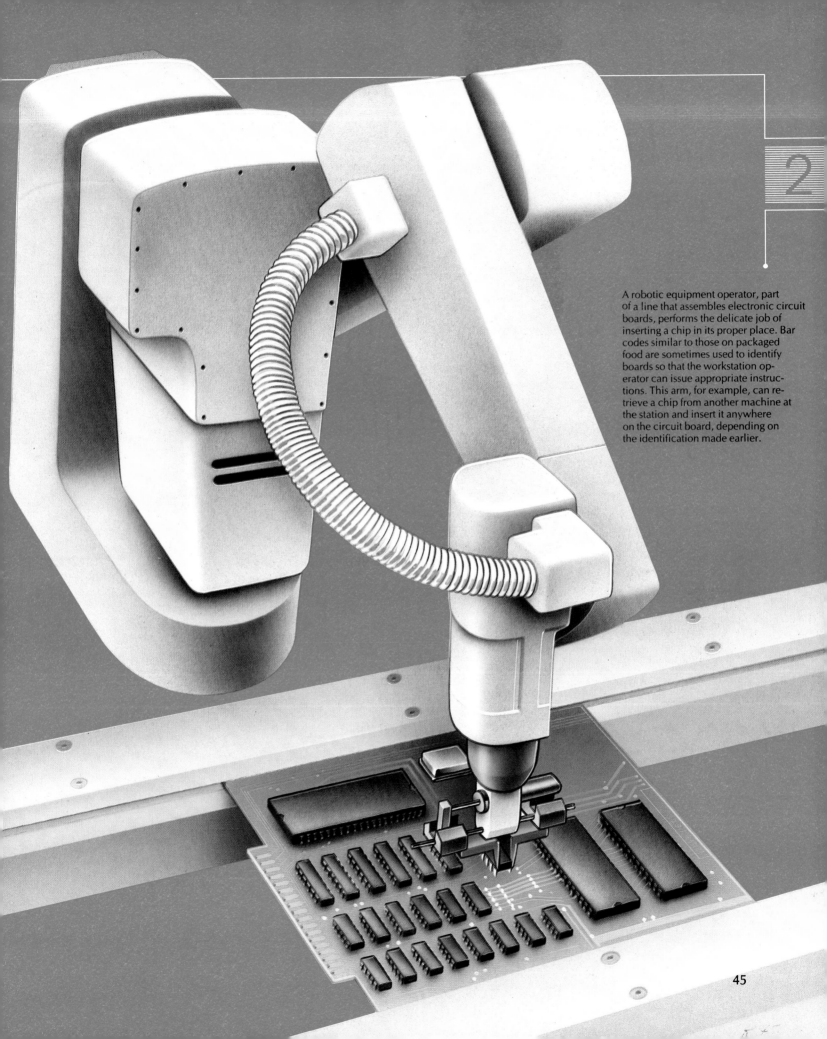

A robotic equipment operator, part of a line that assembles electronic circuit boards, performs the delicate job of inserting a chip in its proper place. Bar codes similar to those on packaged food are sometimes used to identify boards so that the workstation operator can issue appropriate instructions. This arm, for example, can retrieve a chip from another machine at the station and insert it anywhere on the circuit board, depending on the identification made earlier.

45

expect that the simplest robots — those lacking eyes, ears or other sensing ability — could, over a period of 20 years, replace about one million workers, while robots with rudimentary sensing systems could replace three million.

BUILDING A NEW INDUSTRY

The concerns of Ayres and Miller are mitigated by the expectation that robots, while eliminating some human jobs, will create others. One California manufacturer of computer equipment grew from five to 200 employees in just four years after installing 14 robots. The combination of reduced costs and higher quality generated by its robots resulted in such a flood of orders that the human staff had to be enlarged.

Moreover, the robotics industry itself is also counted as a source of employment. The demand for robots is expected to lead to tens of thousands of new manufacturing, programming and maintenance jobs by the turn of the century. "We are creating what is going to be an immense new industry, perhaps as big as the auto industry," says computer scientist Edward Fredkin.

But perhaps the most profound effect of the robotics revolution will be a change in the kind of work that people do. The boring welding, painting and pick-and-place jobs that now burden the workdays of the manufacturing labor force will indeed be gone, assumed by robots. But more and more humans will be required for tasks that machines cannot do. There are some Japanese industrialists, harboring what is probably the most optimistic outlook, who hope that by the year 2000 all their employees will be "knowledge workers," no longer standing on assembly lines but rather sitting at desks and computer terminals to deal with information.

These changes are already under way, and their pace accelerates every year. The steel-collar workers may empty the factory of people, yet they promise to increase the output and vigor of the global economy for the benefit of all.

Robot Arms in Action

Many manufacturing jobs are mind-numbingly repetitive, others are hot and dirty, still others downright dangerous. In some of these situations, industrial robots — usually one-armed machines that are both deaf and blind — can be equipped to perform in place of human workers. But devising robots to carry out even the simplest tasks, such as transferring a part from one place in an assembly line to another, is a considerable challenge.

Robot specialists must draw on the skills and resources of both computer science and mechanical engineering to build systems for industrial arms. The systems must be able to store hundreds of thousands of numbers, reflecting the many positions that the joints of a robotic arm move through in the course of the work. In the arm itself, devices called actuators translate those numbers into elementary movements, while internal sensors detect the movements and report them back to the computer. And all the computations associated with these activities must be accomplished in real time — fast enough to synchronize with their related physical processes.

The following pages trace the essential principles behind the design and operation of an industrial robot, using a spray-painting arm as an example. Spray painting is particularly suited to robotic endeavor since it must be done quickly and accurately: Any imprecision in moving a spray gun over a surface can result in unacceptable flaws — orange-peel rippling, for example, or runs caused by paint build-up — that would require the job to be redone. Moreover, in most factory environments, spray painting is both monotonous and dangerous. In addition to producing harmful vapors in the confines of an industrial paint booth, factory spray painting can be loud enough to damage unprotected ears irreversibly in just eight hours. Measures to shield human workers from chemicals and noise make up much of the cost of industrial paint operations — costs that can be largely eliminated when robots are substituted.

Anatomy of an Industrial Arm

Like the arm of a human factory worker, a robot arm incorporates an articulation system, or skeleton, and a set of muscles, which together function in much the same way as their biological counterparts. The skeleton is made up of rigid links that connect varying numbers of joints, which may slide or rotate. Depending on where in the arm it is located, a joint or a link is known to robotic designers as a shoulder, an upper arm, an elbow, a forearm or a wrist. The robot's muscles consist of actuators that convert hydraulic, electrical or pneumatic energy into power for each joint, often via transmission systems that are analogous to human tendons. An

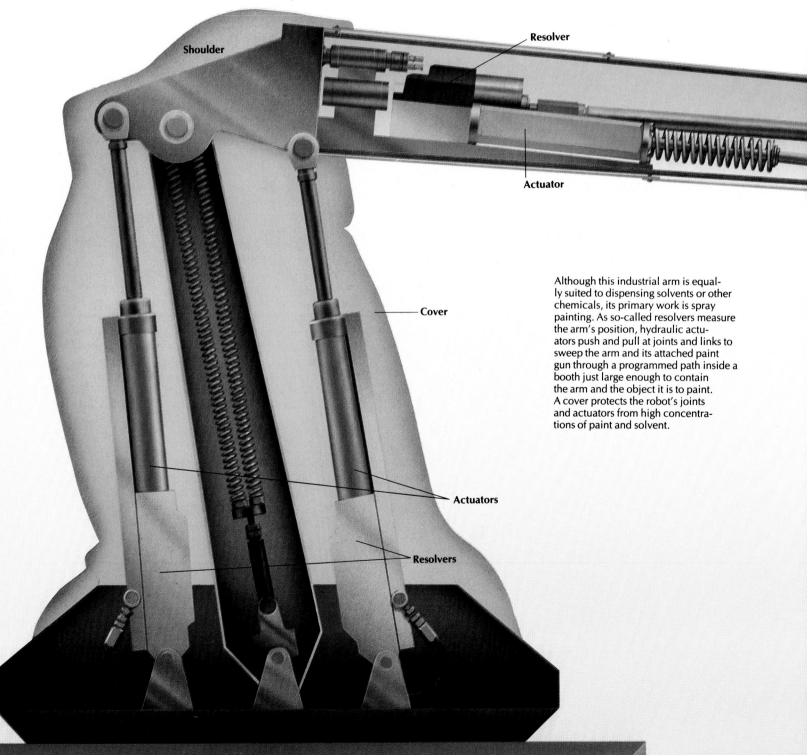

Shoulder

Resolver

Actuator

Cover

Actuators

Resolvers

Although this industrial arm is equally suited to dispensing solvents or other chemicals, its primary work is spray painting. As so-called resolvers measure the arm's position, hydraulic actuators push and pull at joints and links to sweep the arm and its attached paint gun through a programmed path inside a booth just large enough to contain the arm and the object it is to paint. A cover protects the robot's joints and actuators from high concentrations of paint and solvent.

electronic nervous system of wires and sensors carries commands to the muscles and then reports the arm's precise movements back to an external computer.

The similarity between the mechanical and the human worker ends at the robot arm's extremities. Rather than deploying a flexible, many-fingered hand, typical robot arms terminate in special-purpose devices called end-effectors, which are installed directly into the wrist. End-effectors may be changed according to the robot's assigned task. The robot illustrated below and on the following pages, for example, is equipped with a spray gun for painting; the same arm could

be used to apply glue simply by substituting a glue nozzle.

At the opposite end, the arm is bolted to a permanent base. To reduce the number and complexity of computer calculations needed to determine the robot's exact position, the base is generally kept stationary. In a few applications, however, such as spot-welding car bodies, industrial robots are designed to follow an assembly-line item by sliding along a fixed track; others are mounted in overhead tracks called gantries and hang downward from their bases to perform such tasks as assembling automobile engines or drilling holes in a large missile housing.

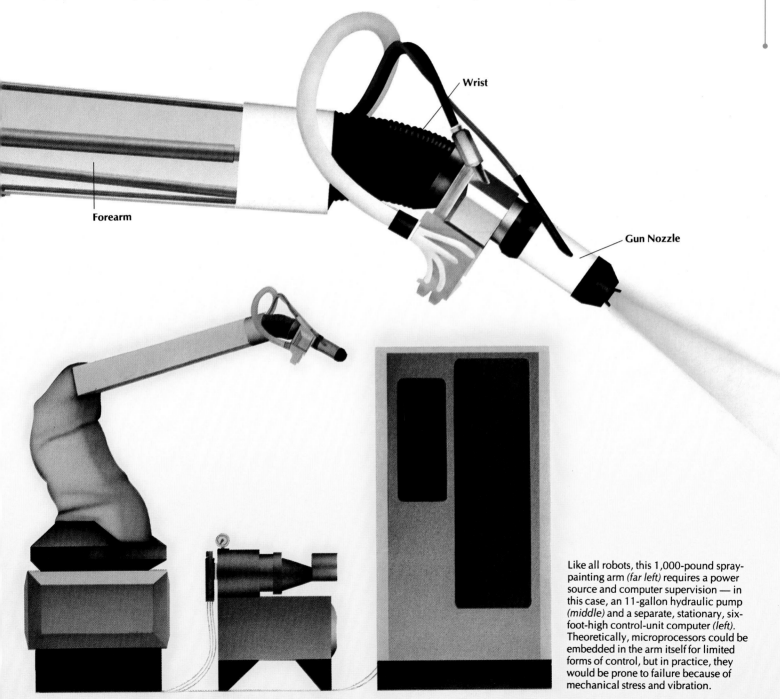

Wrist

Forearm

Gun Nozzle

Like all robots, this 1,000-pound spray-painting arm (far left) requires a power source and computer supervision — in this case, an 11-gallon hydraulic pump (middle) and a separate, stationary, six-foot-high control-unit computer (left). Theoretically, microprocessors could be embedded in the arm itself for limited forms of control, but in practice, they would be prone to failure because of mechanical stress and vibration.

49

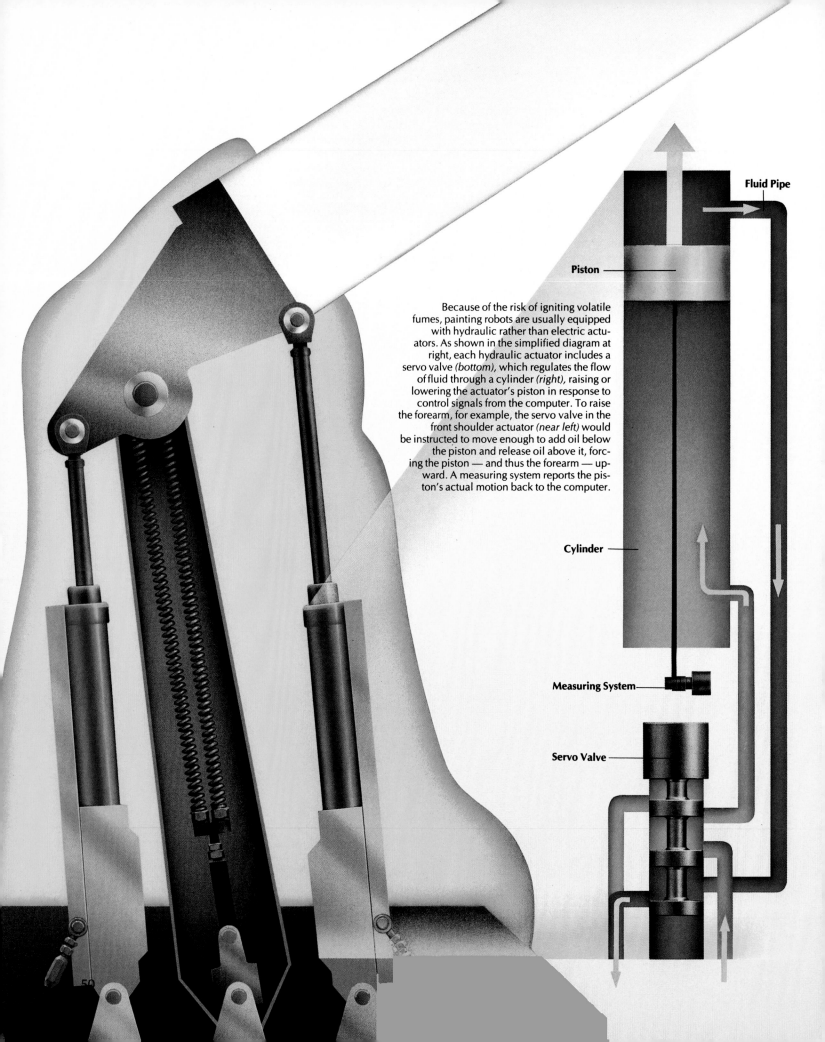

Fluid Pipe

Piston

Because of the risk of igniting volatile fumes, painting robots are usually equipped with hydraulic rather than electric actuators. As shown in the simplified diagram at right, each hydraulic actuator includes a servo valve *(bottom)*, which regulates the flow of fluid through a cylinder *(right)*, raising or lowering the actuator's piston in response to control signals from the computer. To raise the forearm, for example, the servo valve in the front shoulder actuator *(near left)* would be instructed to move enough to add oil below the piston and release oil above it, forcing the piston — and thus the forearm — upward. A measuring system reports the piston's actual motion back to the computer.

Cylinder

Measuring System

Servo Valve

Powerful Muscles under Computer Control

An industrial robot arm's actuators — the muscle power that produces motion — must be carefully regulated to ensure that the arm performs its designated task with the requisite amount of strength and precision. In general, industrial arms make use of either hydraulic or electric actuators, depending on the nature of the robot's job.

Hydraulic actuators *(left)*, which typically use pressurized mineral oil to slide or rotate segments of a robotic arm, are employed in about half of all industrial robots. Hydraulics are several times more powerful than electric motors of the same weight, making them well suited for work in foundries, for example, where they easily manipulate loads of more than 100 pounds. In addition, because they are not subject to electrical arcing, hydraulics are useful in environments where fire is a hazard, such as paint booths or chemical facilities. However, hydraulic systems are prone to drip oil and require frequent maintenance by highly skilled workers.

Electric actuators are most often used for jobs where the movement of hydraulics would be too imprecise — such as inserting components in printed circuit boards. The main drawback to electric motors is their inability to lift heavy objects. Some strength can be gained by employing designs based on the principle of the brushless motor *(below):* By reversing the standard motor design — in which exterior brushes convey current to a rotating interior electromagnet — the brushless motor reduces friction and achieves a higher power-to-weight ratio. But because such motors need electronic rather than mechanical control systems, they are generally more expensive than conventional designs.

A third option, pneumatic drives, is the simplest but least precise power source. The working fluid in a pneumatic drive is compressed air, which is so compressible that unless mechanical stops are used to regulate motion, fine control is difficult.

In a brushless electric motor (also known as an alternating-current servomotor), an external inverter (a special-purpose computer) controls the flow of current to electromagnetic coils spaced throughout the motor's stationary exterior. As the current floods the coils, it creates magnetic fields around them, establishing temporary north and south magnetic poles directly beside their permanent counterparts on the central rotor. Repelled by the exterior poles, the rotor revolves a short distance; to keep the rotor going, the inverter shifts the temporary poles around the stator, in effect chasing the rotor and monitoring it with position sensors until the desired rotation *(arrow)* is complete.

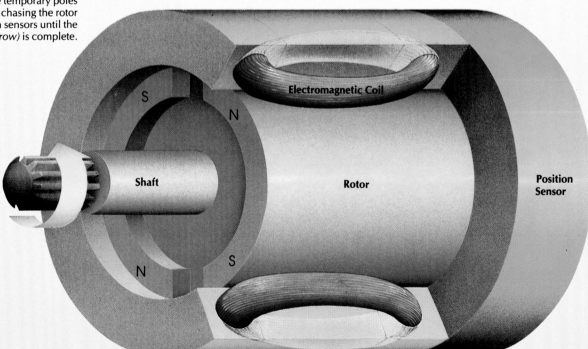

Shaft

S N

N S

Electromagnetic Coil

Rotor

Position Sensor

A Robot's Work Space

Industrial robot arms come in many sizes and designs but tend to fall into a few categories according to the way their joints move. For example, a robot with three sliding, or linear, joints is known as a rectangular- or Cartesian-coordinate robot, because its joints move along the X, Y and Z axes of Cartesian geometry. The space defined by the arm's range of motion takes the shape of a rectangular solid *(below)*. All of the robot's activities occur within this so-called work envelope.

A rectangular-coordinate robot is capable of limited — but very precise — operations. More flexible, but also less precise, is the jointed-arm, or revolute-coordinate, robot, such as the spray-painting arm at right. Three rotary joints — describing a spherical work envelope *(below)* — let a jointed arm reach virtually any part of its workstation, from almost any angle.

Other types of configurations include cylindrical- and polar-coordinate robots. A cylindrical robot arm has two linear joints, allowing the arm to move up and down and telescope in and out; the robot also has a single rotary joint that lets it swivel. A polar robot resembles a piece of field artillery, pivoting on a base and tilting through different elevations to describe a hemispherical work envelope; the arm can also telescope. SCARA robots (for "selective compliance assembly robot arm") are a popular variation on the polar robot's combination of one linear and two rotary joints. In a SCARA design, both rotary joints operate in horizontal planes, yielding precision to within .0005 inch for such tasks as inserting circuit-board components.

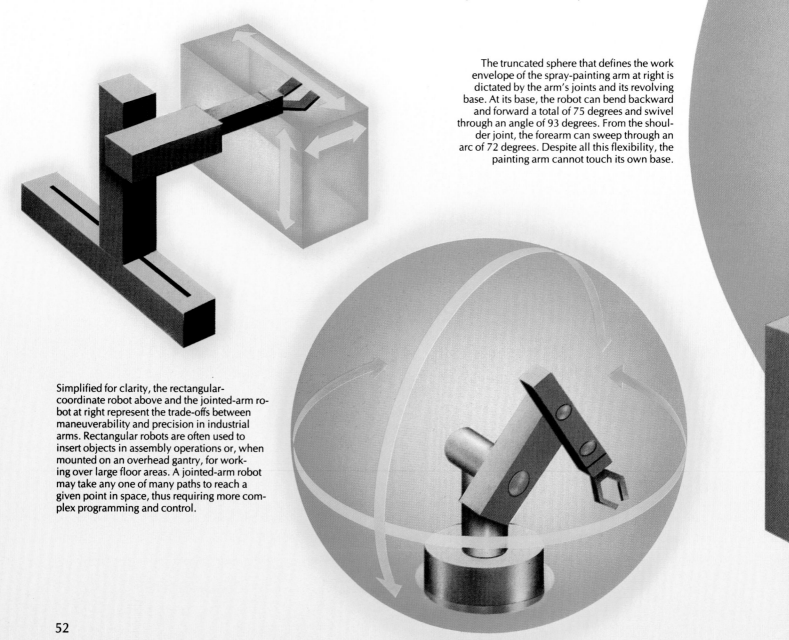

The truncated sphere that defines the work envelope of the spray-painting arm at right is dictated by the arm's joints and its revolving base. At its base, the robot can bend backward and forward a total of 75 degrees and swivel through an angle of 93 degrees. From the shoulder joint, the forearm can sweep through an arc of 72 degrees. Despite all this flexibility, the painting arm cannot touch its own base.

Simplified for clarity, the rectangular-coordinate robot above and the jointed-arm robot at right represent the trade-offs between maneuverability and precision in industrial arms. Rectangular robots are often used to insert objects in assembly operations or, when mounted on an overhead gantry, for working over large floor areas. A jointed-arm robot may take any one of many paths to reach a given point in space, thus requiring more complex programming and control.

2

Training a Robot to Do Its Job

The method of teaching a robot arm how to do its job depends on the complexity of the task and the type of control it requires. Pick-and-place robots, for example, limited to transferring objects from one place to another, are generally con-trolled by two mechanical stops on each axis; the joint can move only between the stops and cannot halt at intermediate points. Programming such an arm is a matter of placing the stops and determining the sequence of action for each axis. In contrast, for robot arms capable of traveling through a set of many specified points, programming is often done by remote control with a so-called teach pendant (right).

But the most complex control is required for the kind of movement characteristic of spray painting. To paint an object of many contours, such as an automobile — or, as shown here

To train a robot for its assigned task, the professional spray painter in this example manually takes the arm through the act of painting a wooden carrousel horse. Internal sensors at each of the robot's joints record its motions 80 times a second for storage by the control unit. Later, the arm can, in effect, play back the series of joint positions, moving at the rate of about 900 millimeters per second and maintaining a distance of about 300 millimeters from the surface of the horse.

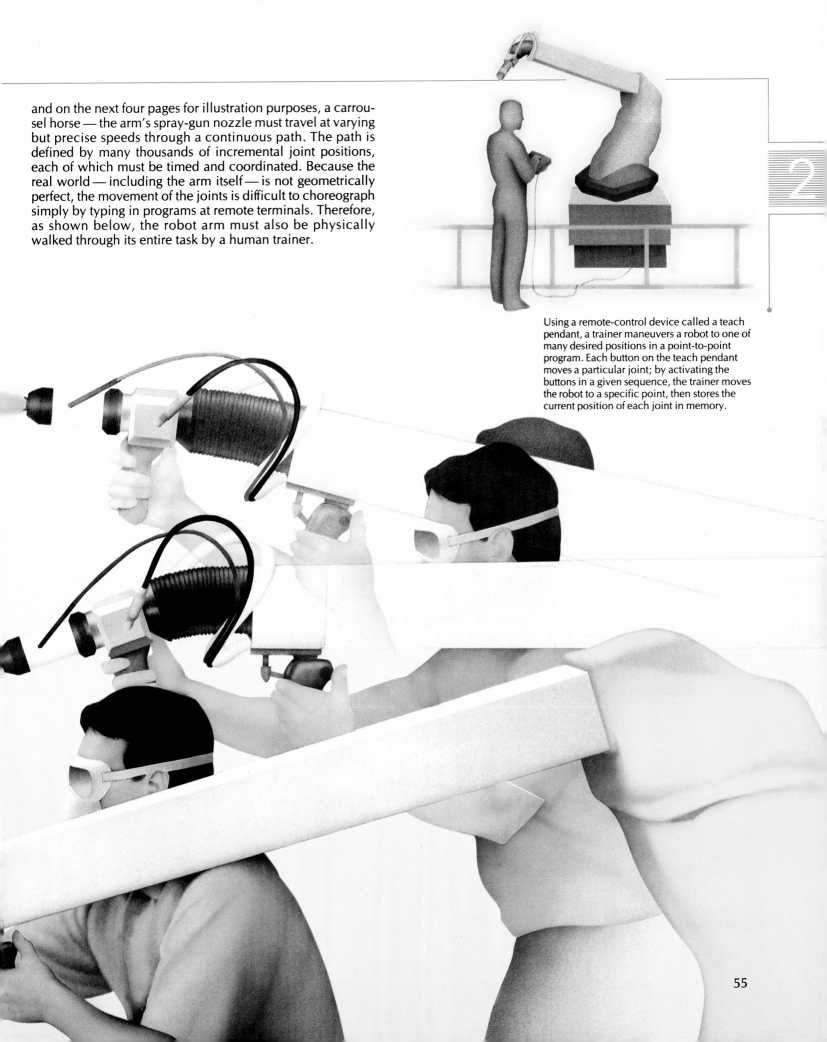

and on the next four pages for illustration purposes, a carrousel horse — the arm's spray-gun nozzle must travel at varying but precise speeds through a continuous path. The path is defined by many thousands of incremental joint positions, each of which must be timed and coordinated. Because the real world — including the arm itself — is not geometrically perfect, the movement of the joints is difficult to choreograph simply by typing in programs at remote terminals. Therefore, as shown below, the robot arm must also be physically walked through its entire task by a human trainer.

Using a remote-control device called a teach pendant, a trainer maneuvers a robot to one of many desired positions in a point-to-point program. Each button on the teach pendant moves a particular joint; by activating the buttons in a given sequence, the trainer moves the robot to a specific point, then stores the current position of each joint in memory.

Feedback for Accuracy

In most cases, industrial robots operate without having the benefit of any of the five senses that human workers depend on to tell them about the external world. Yet the robots are able to move objects unerringly from place to place or, as shown in this fanciful example, to maneuver adequately to paint the front halves of an assembly line of carrousel horses. The secret of this ability can be found in a concept variously

56

known as feedback, closed-loop operation or servo control.

A robot such as this spray-painting arm is not capable of perceiving sensations that originate outside the robot itself. But at every joint, the arm is equipped with internal sensors; these may be optical encoders or, as in the case of the painting arm, electromagnetic resolvers *(box, below)*. The sensors detect a joint's precise position by electromechanical means, translate it into digital form and feed the information back to the robot's control unit. (Open-loop robots, in contrast, do not possess even internal sensors; without feedback, they simply bump from one set of external stops to another.)

The control unit compares information from the sensors with its stored record of joint positions, noting where the arm is supposed to be and where it has actually moved. Almost always there are small discrepancies between the two sets of joint positions, the result of unpredictable variations in materials, atmospheric changes and the effects of inertia and friction. Based on these discrepancies, the control unit determines the next set of commands to be sent to the actuators that power each joint.

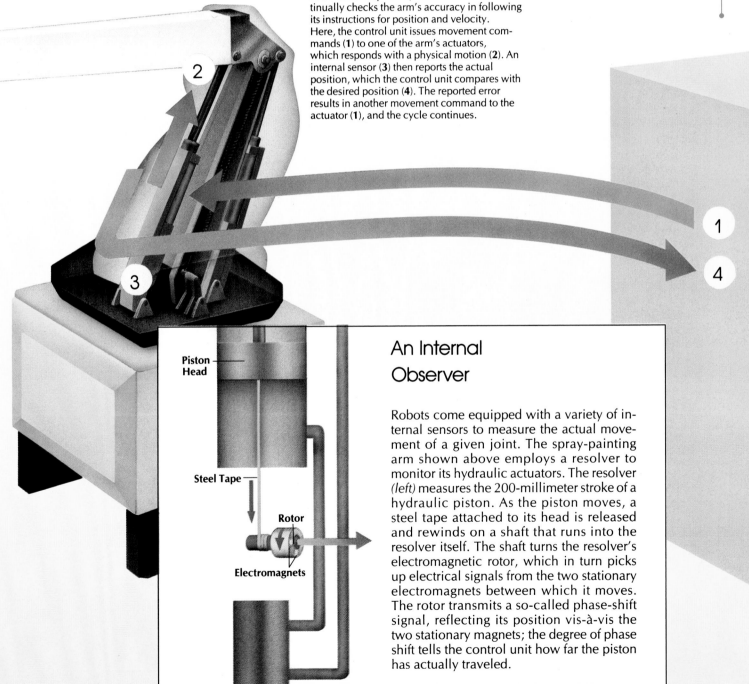

In a closed-loop robot, the control unit continually checks the arm's accuracy in following its instructions for position and velocity. Here, the control unit issues movement commands (**1**) to one of the arm's actuators, which responds with a physical motion (**2**). An internal sensor (**3**) then reports the actual position, which the control unit compares with the desired position (**4**). The reported error results in another movement command to the actuator (**1**), and the cycle continues.

Piston Head

Steel Tape

Rotor

Electromagnets

An Internal Observer

Robots come equipped with a variety of internal sensors to measure the actual movement of a given joint. The spray-painting arm shown above employs a resolver to monitor its hydraulic actuators. The resolver *(left)* measures the 200-millimeter stroke of a hydraulic piston. As the piston moves, a steel tape attached to its head is released and rewinds on a shaft that runs into the resolver itself. The shaft turns the resolver's electromagnetic rotor, which in turn picks up electrical signals from the two stationary electromagnets between which it moves. The rotor transmits a so-called phase-shift signal, reflecting its position vis-à-vis the two stationary magnets; the degree of phase shift tells the control unit how far the piston has actually traveled.

A Need for Teamwork

In many industrial situations, a solitary arm simply is not adequate. Assembly operations, for example, often require that each of two parts be repositioned as they are fitted together, an effort best performed by two arms. This kind of coordination is natural to a human worker or a pair of workers, but with robot arms it adds considerably to the programming challenge. For example, the program must be able to compensate for the force exerted by each arm, whether they are holding two parts that must be assembled or jointly handling the same object. Similarly, the arms must be able to adjust to each other's speed. But perhaps the most visible problem of coordinating multi-armed robots is that of collision avoidance.

As depicted in the imaginary task shown here — attaching a unicorn's horn to the forehead of a carrousel horse — teamwork between two robotic arms means that they must enter each other's work envelope. One approach to the collision problem is to link the arms through a central microcomputer, so that each must, in effect, ask permission to enter the other's work envelope. In one very advanced system used by General Motors, pairs of robot arms work together to paint the insides of automobiles. One arm opens the car door, then moves aside while the partner sprays the car's interior; when the partner withdraws, the first arm closes the door.

Theoretically, it should also be possible to program a computer to observe the arms' movements. Initially, each arm would be put through its paces to establish a so-called movement signature; should the computer find that an arm's actual motion varies too much from its signature and threatens to cause a collision, the computer would call a halt. A third possibility would be to use individual external sensors rather than linking the arms with a central computer. Each arm would possess an acoustical sensor (similar to sonar) for measuring the distance to nearby objects. If other robots came too close, the robot arm would immediately stop moving; after a predetermined delay, the arm would start up again slowly and cautiously, making sure that the obstacle had gone.

In this fanciful example of the coordination of a pair of industrial robots, an arm with a glue nozzle cooperates with a gripper-equipped robot to insert a unicorn's horn in the forehead of a carrousel horse. By communicating through a common control unit, the robots not only avoid colliding with each other but adjust their individual pace to accommodate that of their partner — waiting to insert the horn, for example, until the glue arm has withdrawn.

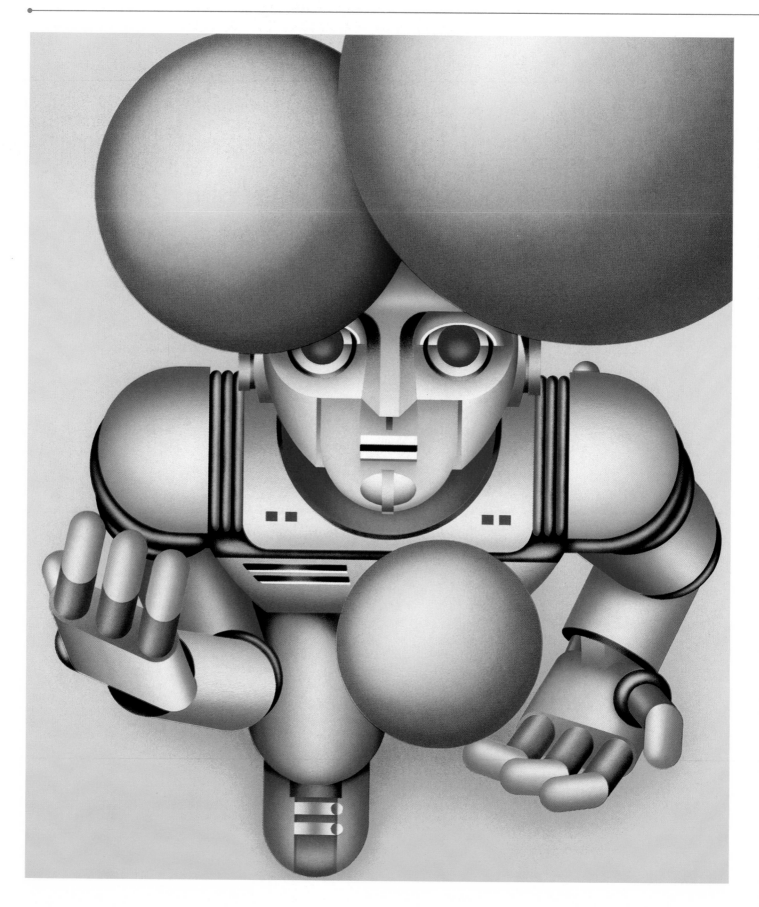

New Talents on the Drawing Board

In late 1983, a dancer unlike any ever seen on earth began to perform in the laboratory of Marc H. Raibert, a professor of computer science at Carnegie Mellon in Pittsburgh. The terpsichorean was a machine, a homely, three-foot-tall arrangement of tubing, sensors and hydraulic valves mounted atop a slender, telescoping metal cylinder. In repose, it gave no hint of its abilities. But once the power was turned on and the contraption came to life, hopping about with all the agility and vigor of a kangaroo, its identity was unmistakable. The little creature was clearly a kind of automated pogo stick.

No witness of one of its performances could fail to be amazed. Energized by an umbilical cord that carried computer data, compressed air and pressurized oil, Raibert's creation would bound across the floor at just under five miles per hour, each stride nearly two feet long. The compressed air provided thrust to the telescoping leg, and two hydraulic actuators caused the leg to pivot so that it met the floor at exactly the right angle after each hop. Gyroscopes and other sensors kept precise track of critical factors — the leg's length, its motion relative to the body of the hopper, the force of the foot hitting the ground and the body's position. At the other end of the umbilical cord, a computer worked out all the adjustments the hopper would have to make in order to stay upright. The mathematics were surprisingly simple, dividing naturally into three parts — computations dealing with the height of the hop, the position of the foot at landing and the attitude of the hopper's body.

As the machine carried out its acrobatics, it would cant, wobble and sometimes seem ready to crash to the concrete floor, but it would always catch itself and stay upright, if not precisely vertical. Sometimes Raibert gave it a push as it passed. Although this would cause it to lurch drunkenly, the little leaper would display a gymnast's powers of recovery, regaining its balance with a series of stationary hops, then continuing to jump around the room.

For all its entertaining ways, the computer-controlled pogo stick was part of a sober-minded scientific quest. Raibert, who left Carnegie Mellon for M.I.T. in 1986, wants ultimately to create "dynamic legged systems"—robots that walk much the way a horse or a human being does. He designed the hopping machine in order to learn more about the phenomenon of balance, reasoning that if he could get a machine to hop on one leg, he could isolate balance as a factor without having to worry at the same time about coordinating the movements of several legs. Pogo sticks, he says, "are springy; they interact with the ground. There seems to be something fundamentally similar about the way pogo sticks behave and the way legs behave."

He and the handful of scientists working at the far limits of robotics research tend to think about the applications of their findings only in general terms. Their research is an end in itself; they have little in common with the engineers who design the current generation of industrial robots. Despite such robots' prowess at assembling, drilling, welding or spray painting, what they can do pales beside what they cannot. Bolted to the shop floor, operating in an environment that may have special lighting or backgrounds to simplify visual perception, selecting from a sharply limited repertoire of actions, the typical industrial robot of today is far from resourceful.

But robotics, as Raibert puts it, "doesn't mean making machines produce things in the factory." He and other scientists at research centers such as Carnegie Mellon, Stanford University's Artificial Intelligence Laboratory and the Massachusetts Institute of Technology are inclined to see the future of robotics in terms of mobile, general-purpose machines that can flourish in an unstructured environment. In various ways, they dream of a robot that will walk — perhaps even run — on leglike appendages. It will guide itself as it moves, recognizing what it seeks or needs to avoid. Vision and other sensing systems will let it identify objects in its world, and it will manipulate them with human-like dexterity. It will have the potential, in short, to be the helpmate to humankind long envisioned by science fiction writers. It may work in a hazardous mine or tunnel; it may take over many of the underwater jobs currently done by divers or by manned and remotely controlled submersibles; it may fight forest fires, perform construction or farm work, or act as tomorrow's astronauts; it may have military roles to play, standing sentry duty indefatigably or defusing a bomb with a locksmith's touch.

Few researchers on the frontiers of robotics feel able to work on more than one kind of problem at a time — mobility, perhaps, or sensing, or the development of general-purpose hands. At this stage, they are beginning to comprehend nature just barely well enough to build costly mechanical and electronic parodies of brain, leg, hand and eye. And beyond nuts-and-bolts difficulties lies an Everest of programming conundrums. According to James S. Albus, head of the robot-systems division at the National Institute of Standards and Technology, "The software does not exist at any price that could control a walking machine with two arms and a full set of force, touch and vision senses in the performance of tasks like building a brick fireplace, laying a hardwood floor, installing a bathtub or painting the front of a house."

EXAMINING BASIC PROCESSES
Most researchers at the outer reaches of robotics, in fact, feel that while scenarios involving future robotic longshoremen, vehicles and army sentries make for entertaining dinner chitchat and worthy long-range goals, the real value of research robots is that they help to explain the processes that underlie motion, touch and other interactions with the physical world. "We're trying to understand those principles, and that's exciting," says Raibert.

Of the various challenges facing these researchers, creating a robot that simply propels itself along presumably should be the easiest. But the most useful form of locomotion promises to be walking, and that is a difficult proposition indeed.

The virtues of walking as a means of getting around are clear enough. About half the land on the earth's surface is impassable to wheeled and tracked vehicles. Wheels are particularly handicapped. They need continuous support, whereas legs can exploit isolated footholds. On soft ground, a wheel wastes energy by having to constantly climb out of a depression it has formed; a leg, by contrast, pushes the soft soil or sand downward, compacting it in the process and thus helping itself along. And a legged vehicle, unlike a wheeled one, could step over obstacles and walk along rough or uneven terrain while the body and cargo traveled smoothly. Tracked vehicles do better than wheels on soft or rough ground, but a legged vehicle could cope with far more of the earth's terrain — at least in theory.

One of the first mechanical walkers appeared in 1870, when a "kinematic linkage" designed earlier in the century by the Russian mathematician P. L. Chebyshev found its way into a machine that could step along a path — provided that the path was smooth and horizontal. The machine's linkages could only generate specific, fixed stepping patterns. They could not cope with bumps and slopes. Scientists continued to investigate various linkage designs in the decades that followed. But their efforts only underscored the point that, to be useful, walking machines had to be able to deal with unpredictable terrain by adjusting to the footholds available. And that meant that someone, or something, had to control the legs. Ralph S. Mosher decided to harness a human.

A WALKING TRUCK

Mosher was an engineer with General Electric in 1962, when the Advanced Research Projects Agency of the Defense Department began to look for a rough-ground transport vehicle that would get about by walking. GE originally planned to build a two-legged, upright walking machine several times the height of a person. A driver sitting inside would control the machine through levers attached to servomechanisms that amplified his movements while telegraphing back a sense of what the machine was doing. But project leaders ultimately decided that a quadruped design would be more practical. The result was the "walking truck," as GE called it. The device stood 11 feet tall, weighed one and a half tons and had a 90-horsepower gasoline engine. Each of the driver's limbs controlled a lever or pedal that, in turn, controlled one leg by means of hydraulic actuators. The machine was powerful and versatile. It could stride along at about five miles per hour, navigate a stack of railroad ties and push a jeep out of the mud. Unfortunately, it also reduced its operators — even Mosher, who trained rigorously — to panting exhaustion in 10 minutes, for they had to manually track and coordinate the movements of three joints in each leg, or 12 in all.

GE's walking truck was an inspiration to Robert B. McGhee, then a specialist in guided missiles at the University of Southern California. His interests had gradually been shifting toward human motion, with the aim of improving prosthetic limbs for the handicapped. Prosthetics led him to robotics. When McGhee heard about GE's vehicle, he concluded that he could get around the impracticality of human control by building a mechanical quadruped with legs coordinated by a computer rather than a human.

To do that, McGhee first needed to break down the walk of a four-legged beast into discrete elements. He took photographs of a horse walking and studied them for hours — learning, for example, that the leg has to bend at the knee before it can swing forward. Then McGhee and Andrew Frank, one of his students, went to

Deciphering
the Mysteries
of Speech

Teaching a machine to understand human speech is one of the most difficult challenges facing roboticists. Natural spoken language — a code that human children quickly break — is nothing less than a thicket of traps and uncertainties for machine speech-recognition systems. Irrelevant sounds — such as paper crumpling or an intermittent "ah" or "you know" — must be sifted out. Many words, such as "reel" and "real," sound alike, and phrases may run together so that "meet her" becomes "meter." Depending on the accent of the speaker, the same word may be pronounced several ways; in some sections of the United States, for example, the word "school" sounds much like "skull."

Despite these difficulties, scientists are beginning to lay the groundwork for spoken communication between robots and humans. Already, a few rudimentary speech-recognition systems are used for tasks such as taking inventory and handling airline baggage. These systems operate in a very constrained environment. Their vocabularies are limited to about 100 task-specific words, compared with a normal human vocabulary of roughly 50,000 words. To avoid problems arising from differences in pronunciation, such systems are trained, in effect, to recognize the voices of only a few speakers — who must in turn be trained to enunciate clearly and to pause between words.

The sound waves produced by single words spoken into a microphone are translated into electrical signals; these, in turn, are digitized and converted into an array of numbers for the robot's computer. The array may also be converted into an image called a spectrogram, like the one shown at right, which reveals the distinctive pattern formed by spoken words. The system compares incoming signals with stored arrays, or templates. If the incoming array matches a template, within acceptable tolerances, the robot recognizes the word and acts accordingly.

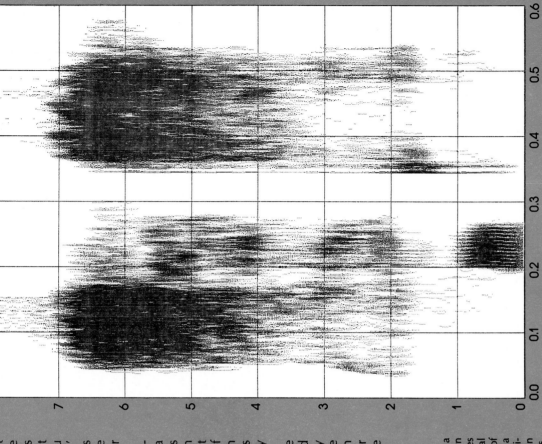

The spoken word "six" generates a distinctive pattern of frequencies, shown here in a spectrogram. Frequencies range from low to high on the vertical axis, plotted over time — in tenths of a second — on the horizontal axis. In a speech-recognition system, the variations in frequency are converted to an array of binary ones and zeros.

FIVE SIX SEVEN

In order to recognize an incoming set of frequencies as a given word, the computer compares it with stored templates (above), one for each word in its vocabulary. (For clarity, templates are represented here as simplified spectrograms rather than as arrays of numbers.) Because words are never said twice in exactly the same way, the match is never perfect, and the computer must look for the best approximation. With the aid of a process called dynamic time warping, the computer manipulates the template until it fits the incoming pattern.

A Focus on the Building Blocks

Speech-recognition systems designed to match whole words with templates are severely handicapped in the real world. Because such systems must compare all incoming speech signals with all stored templates, they quickly run into what one expert has called "the wall of vocabulary size." The larger the desired vocabulary, the more storage capacity is required for templates and the longer it takes the computer to find the correct match for an incoming signal—until the system is overwhelmed and simply ceases to function usefully.

In an effort to get around these problems, researchers have borrowed a notion from linguistics—opening up words to examine the basic sounds from which they are constructed. Each word in the English language is composed of one or more of 40 acoustic building blocks, called pho-

nemes. An s sound, for example, when studied in a spectrogram, always appears as a high-frequency, dark, rectangular shape. Systems based on such isolated features can thus recognize words and key phrases as strings of phonemes, an approach that makes fewer demands on memory and processing time.

With feature-based systems, as they are called, designers can also avoid some of the difficulties posed by differences in regional pronunciation (below, left) and by the blurring between words—known as coarticulation—that occurs when people speak (below). As illustrated here, feature-based systems deal with these effects by focusing on the phonemes that remain recognizable from one speaker to another and in virtually any context.

The Keys to Dialect

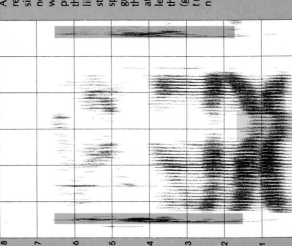

A feature-based system reads incoming speech signals as strings of phonemes rather than as whole words. The computer searches only for those parts of the signal likely to remain constant, no matter who is speaking. In the spectrograms shown here are three different pronunciations of the word "dialect." In all three cases, the patterns for d (green), l (yellow) and t (purple) are recognizably distinct.

When Words Blur

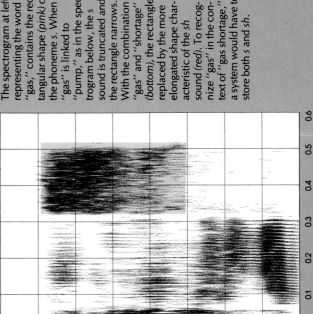

The spectrogram at left, representing the word "gas," contains the rectangular shape (pink) of the phoneme s. When "gas" is linked to "pump," as in the spectrogram below, the s sound is truncated and the rectangle narrows. With the combination of "gas" and "shortage" (bottom), the rectangle is replaced by the more elongated shape characteristic of the sh sound (red). To recognize "gas" in the context of "gas shortage," a system would have to store both s and sh.

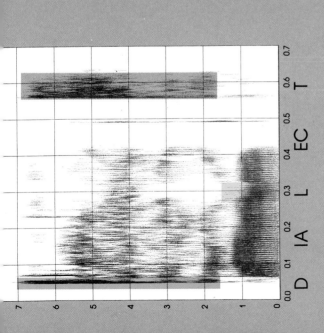

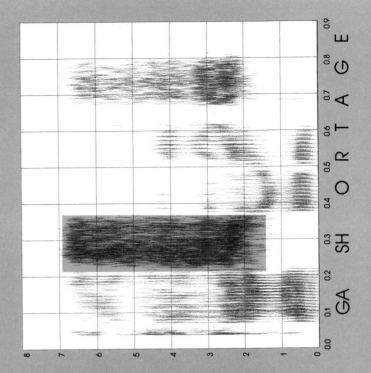

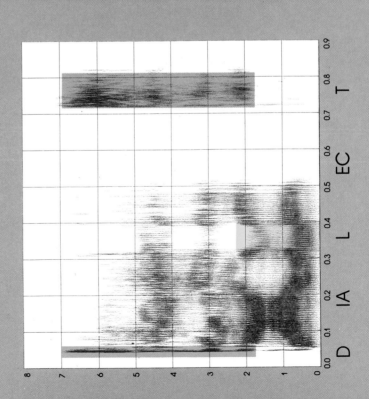

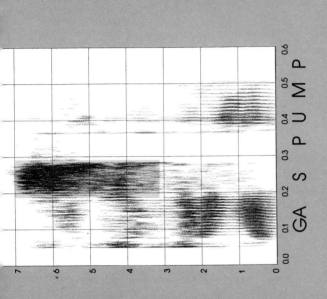

work. In 1966, the two unveiled a device they called the Phony Pony. It looked and behaved much like a table frame come to life. Equipped with an electric motor at each of its eight joints — one at each leg's "hip" and "knee" — it inched forward, a leg at a time. The movements were "very jerky," said McGhee, "like a pantomime of a robot."

Though graceless and slow, the Phony Pony was a considerable accomplishment, because it demonstrated that robotic vehicles did not have to be exclusively controlled by a human operator. In 1968, McGhee moved to Ohio State University, where he continued his work on robotic walkers. Nine years later, he showed off a new design, known as the OSU Hexapod. It was a six-legged, insect-like vehicle about four feet long and almost five feet wide. Linked by cables to a computer and a source of electrical power for the motors that drove its legs, it lumbered along at about a quarter of a mile an hour, climbing over obstacles and up shallow stairs. Not everyone was impressed. Senator William Proxmire, who has frequently been skeptical or even scornful of scientific research projects funded with government money, labeled it the "bionic bug." He added it to the long list of recipients of an award he had dreamed up to express his derision: the "Golden Fleece."

But the Hexapod was an effective working laboratory for McGhee and co-worker Kenneth J. Waldron, a professor of mechanical engineering at Ohio State, and they soon went on to even more ambitious designs. Late in 1985, another creation — a six-legged aluminum behemoth 17 feet long and 10 feet high called the Adaptive Suspension Vehicle — took a walk. Developed with Defense Department money, the ASV has come closer to the notion of a sure-footed, muscular military mule than any other machine to date. It was built to travel, with a human aboard to do some of the controlling, at about eight miles per hour; to step over a ditch nine feet wide; to surmount a seven-foot-high obstacle; and to climb a 60-percent slope. All this is accomplished with five million dollars' worth of machinery.

CONTROLS FOR THREE TONS OF ALUMINUM
Unlike the Hexapod, the ASV has onboard computers — 17 powerful microprocessors — and its own power supply: a 70-horsepower engine that drives a flywheel at up to 12,000 revolutions per minute. The flywheel, in turn, drives pumps that energize hydraulic actuators at each of the leg joints. The operator controls the three-ton robot with a joystick. Moving the stick forward or back causes the machine to proceed in the indicated direction at various speeds. If the stick is pushed to the side, the ASV will move crabwise. Twisting the stick prompts the robot vehicle to turn in place. And buttons on the joystick control the attitude of the body and its height above the ground. All of the control options can be combined as the terrain demands.

Perhaps more ambitious than either the Hexapod or the ASV is NASA's unmanned Planetary Rover. Intended for exploring Mars and other nearby worlds, this six-legged walking machine is expected to take its first strides as a prototype by the mid-1990s. Unlike the Hexapod, the Rover will rely entirely on computers for guidance. They are to be installed, along with cameras and other sensors, on a platform atop the vehicle's limbs, whose nine-foot length permits the Rover to step over obstacles as tall as three feet. Pressure-sensitive foot pads prevent missteps onto soft ground that could cause the walker to topple.

Walkers such as the Planetary Rover belong to a category of machines that roboticists call statically stable. Most such machines are designed to maintain their balance by keeping at least a tripod of feet on the ground at all times — although two-legged versions are possible, provided the feet are big and the transition of stepping from one foot to the other is carefully managed. Work on statically stable walking machines is being pursued by roboticists around the world — most notably in the Soviet Union and Japan. In the mid-1970s, for example, researchers at the University of Moscow built a six-legged walking machine that kept its body level with the aid of sensors analogous to those in the human inner ear. A series of other six-legged projects followed.

Japanese interest in walking machines also began in the 1970s. One early project was a two-legged, hydraulically propelled robot named Wabot 1, built by Ichiro Kato of Waseda University. In the late 1980s, Japan's Advanced Robot Technology Research Association (ARTRA) began work on a pair of more practical walking machines. The first, a quadruped with two arms, is intended for maintenance and inspection duties in a nuclear power plant. The second, with six legs, will be a firefighter that can withstand temperatures of 1,450° F and see through flames and smoke with ultrasonic and infrared-laser rangefinders.

A BETTER BALANCING ACT
In the matter of locomotion, statically stable walkers may not be the wave of the future. Marc Raibert makes a strong case for developing walkers that have an active sort of balance — technically called dynamic stability. This simply means that a machine can stay upright without having its center of gravity directly over the supporting legs at all times. The locomotion of a kangaroo, dog or human exemplifies dynamic stability in action. Clearly, robot walkers would be far more versatile if they had such balancing skills. Raibert enumerates the advantages of dynamic stability this way: "If a legged system can tolerate tipping, then it can position its feet from the center of gravity in order to use widely separated or erratically placed footholds. If it can remain upright with a small base of support, then it can travel where obstructions are closely spaced or where the path of firm support is narrow. The ability to tolerate intermittent support also contributes to mobility by allowing a system to move all its legs to new footholds at one time, to jump onto or over obstacles, and to use short periods of ballistic flight for increased speed. Animals routinely exploit active balance to travel quickly on difficult terrain; legged vehicles will have to balance actively, too, if they are to move with animal-like mobility and speed."

Raibert himself is energetically chasing that vision. Having worked out some of the basic problems of dynamic stability with his pogo-stick hopper, he has gone on to design biped and quadruped machines that can balance dynamically.

The mathematical relationships that controlled the balance of his one-legged machine have turned out to adapt nicely to a biped. The reason: With a biped that runs like a person, only one leg is working at any one time. So long as a computer keeps track of which leg is active and makes sure that the other one is out of the way, a one-leg mathematical approach can regulate each of the biped's legs. His first biped has run at 11 miles per hour — more than twice the hopper's speed. His quadruped, too, can be controlled by the pogo-stick mathematics. Because quadrupeds' legs often work in pairs, each pair can be treated and controlled like a single leg; that reduces four legs to the equivalent of two. The machine, so far as its dynamic balancing system is concerned, is a biped, which can be regulated like a one-legged hopper.

These initiatives remain very much at the laboratory stage of exploration. Still, recent progress on the frontier of legged robots has been rapid. Robert McGhee's six-legged Adaptive Suspension Vehicle, for example, has achieved an energy efficiency — measured in terms of miles per gallon per ton — that is comparable to tracked vehicles'. Further improvements can be expected, because walking-machine technology is new and also because nature's walkers show that far higher levels of efficiency are physically possible: Animals and humans expend a tenth as much energy as the ASV when they walk.

MOBILE ROBOTS ON THEIR OWN

Given a robot that can propel itself effectively, a logical next step would be to endow it with the talent to guide itself — a degree of autonomy, in other words. In the early 1960s, Johns Hopkins University was the scene of an experiment along those lines: Researchers designed a squat, tub-shaped mobile robot that, without human guidance, roamed the halls in quest of electrical "food." Called the Beast, it kept its distance from the walls by means of an ultrasonic range sensor. The robot was also equipped with a photocell optical system that looked for the distinctive black rectangular cover of an electrical wall outlet. Finding one, it would plug in a special arm and recharge its batteries before continuing on its way.

The Beast possessed some basic processing skills that allowed it to respond to input from its sensors, but its thinking powers were dim indeed. In 1969, however, a robot with far greater computational abilities made its debut. This was Shakey, developed by the Stanford Research Institute (now SRI International). Built for research in artificial intelligence, it solved problems in logic by processing information gathered through a video camera. Shakey, so named for the unsteadiness of its movements, wobbled around its confined laboratory on a set of small drive and caster wheels, avoiding obstacles or pushing and arranging a set of blocks according to instructions. The robot achieved its goals with the help of STRIPS, a powerful computer program that enabled it to plan a sequence of actions. On one celebrated occasion, Shakey was told to push a block off a raised platform. First it found a wedge-shaped object and shoved it against the platform. Then it maneuvered up this improvised ramp and accomplished its mission.

Able to interact independently with its environment through its own senses and actions, Shakey was a robot in the fullest meaning of the word. But aside from research, it had no practical use. The SRI robot was extremely slow, taking as much as an hour to locate and identify a block or platform in its uncluttered, bare-walled laboratory. It was intended to function only in this highly restricted setting and could never have navigated in an ordinary room, let alone in surroundings as complex as an urban sidewalk.

One of the prime visionaries in the field of navigating robots is Hans Moravec of Carnegie Mellon. Moravec, says a colleague at the university's Robotics Institute, "is driven by an absolute belief" that machines will evolve in the same way

Mimicking Natural Flexibility

In their efforts to expand the range of robotic applications, researchers are looking beyond traditional designs to examine a variety of potential models from the biological world. An early success along these lines is an industrial arm *(below)* whose sinuous flexibility is patterned after the structure of the vertebrate spine. Two other approaches, shown on the following pages, take as their models the musculature of elephant trunks and the grasping behavior of octopus tentacles.

Researchers have considered a number of possible applications for robots based on biological structures, including wall-climbing, spider-like machines that could carry ropes to people trapped in a burning building, and a very slender, snakelike robot for gastrointestinal examinations. Experimental legged systems often mimic the gaits of animals *(pages 81-89)*, and some computerized vision systems have been inspired by that of the frog.

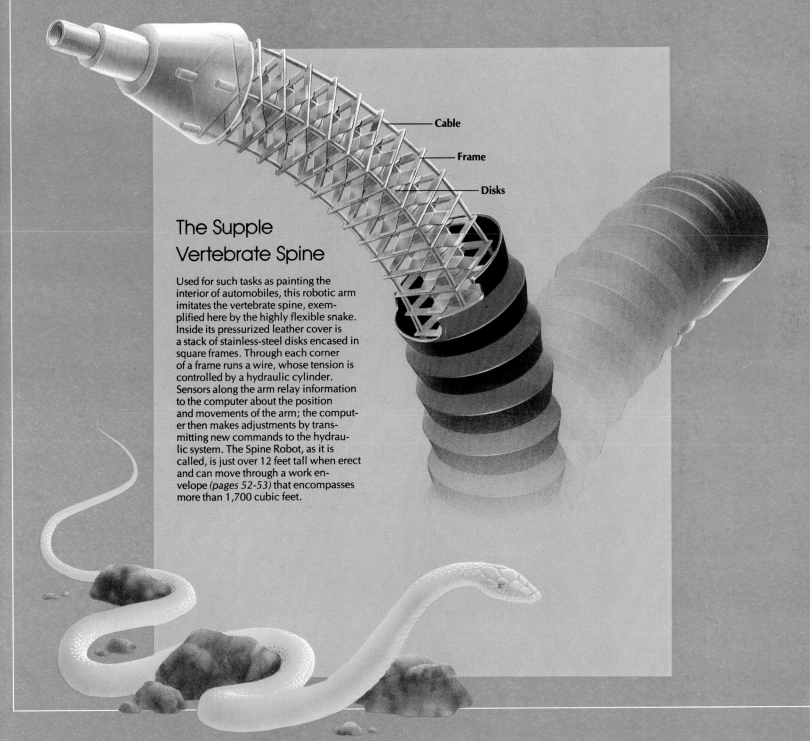

Cable

Frame

Disks

The Supple Vertebrate Spine

Used for such tasks as painting the interior of automobiles, this robotic arm imitates the vertebrate spine, exemplified here by the highly flexible snake. Inside its pressurized leather cover is a stack of stainless-steel disks encased in square frames. Through each corner of a frame runs a wire, whose tension is controlled by a hydraulic cylinder. Sensors along the arm relay information to the computer about the position and movements of the arm; the computer then makes adjustments by transmitting new commands to the hydraulic system. The Spine Robot, as it is called, is just over 12 feet tall when erect and can move through a work envelope *(pages 52-53)* that encompasses more than 1,700 cubic feet.

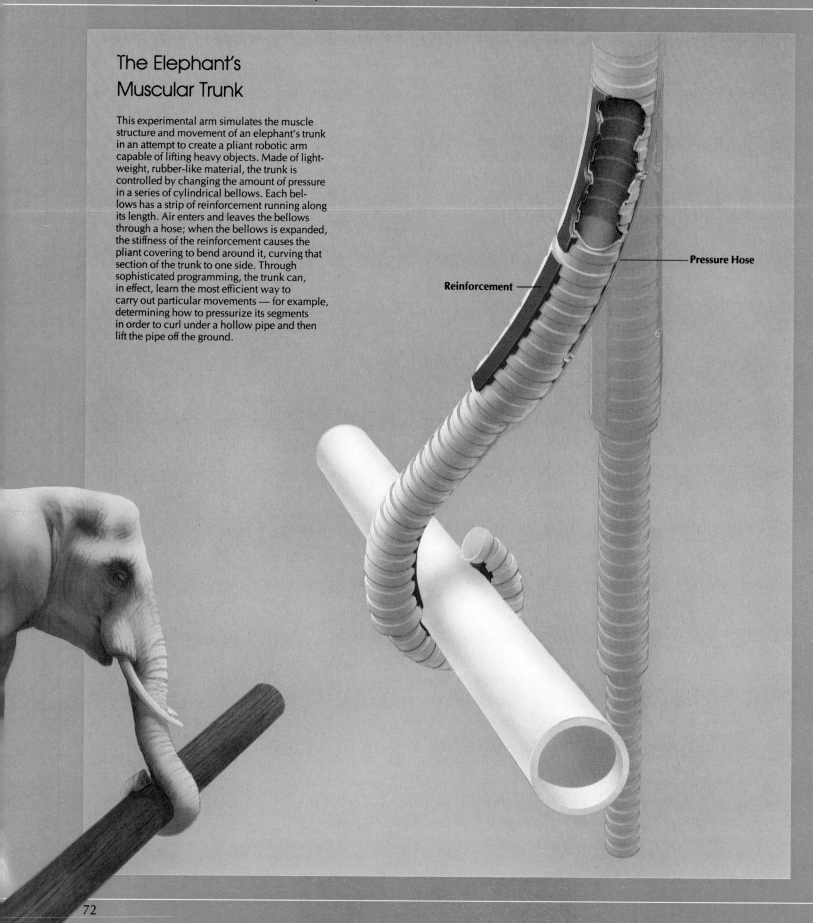

The Elephant's Muscular Trunk

This experimental arm simulates the muscle structure and movement of an elephant's trunk in an attempt to create a pliant robotic arm capable of lifting heavy objects. Made of lightweight, rubber-like material, the trunk is controlled by changing the amount of pressure in a series of cylindrical bellows. Each bellows has a strip of reinforcement running along its length. Air enters and leaves the bellows through a hose; when the bellows is expanded, the stiffness of the reinforcement causes the pliant covering to bend around it, curving that section of the trunk to one side. Through sophisticated programming, the trunk can, in effect, learn the most efficient way to carry out particular movements — for example, determining how to pressurize its segments in order to curl under a hollow pipe and then lift the pipe off the ground.

Pressure Hose

Reinforcement

The Octopus's Adaptable Grip

The two parts of this gripper emulate the flexibility of an octopus. Like their biological counterparts, the robotic tentacles can conform to fragile objects of any shape and hold them with uniform, gentle pressure. Shown here wrapped around a summer squash, the tentacles might be used to pick easily damaged fruit. A variation of the design might handle animals, turn hospital patients in their beds or lift a small child.

A single pair of wires — one for gripping, the other for releasing — runs through both tentacles around a series of pulleys. A motor and a clutch for each wire are located at the base of the tentacles. Engaging the grip clutch pulls the grip wire, and the tentacles contract to wrap around an object; engaging the release clutch pulls the release wire and causes the tentacles to relax.

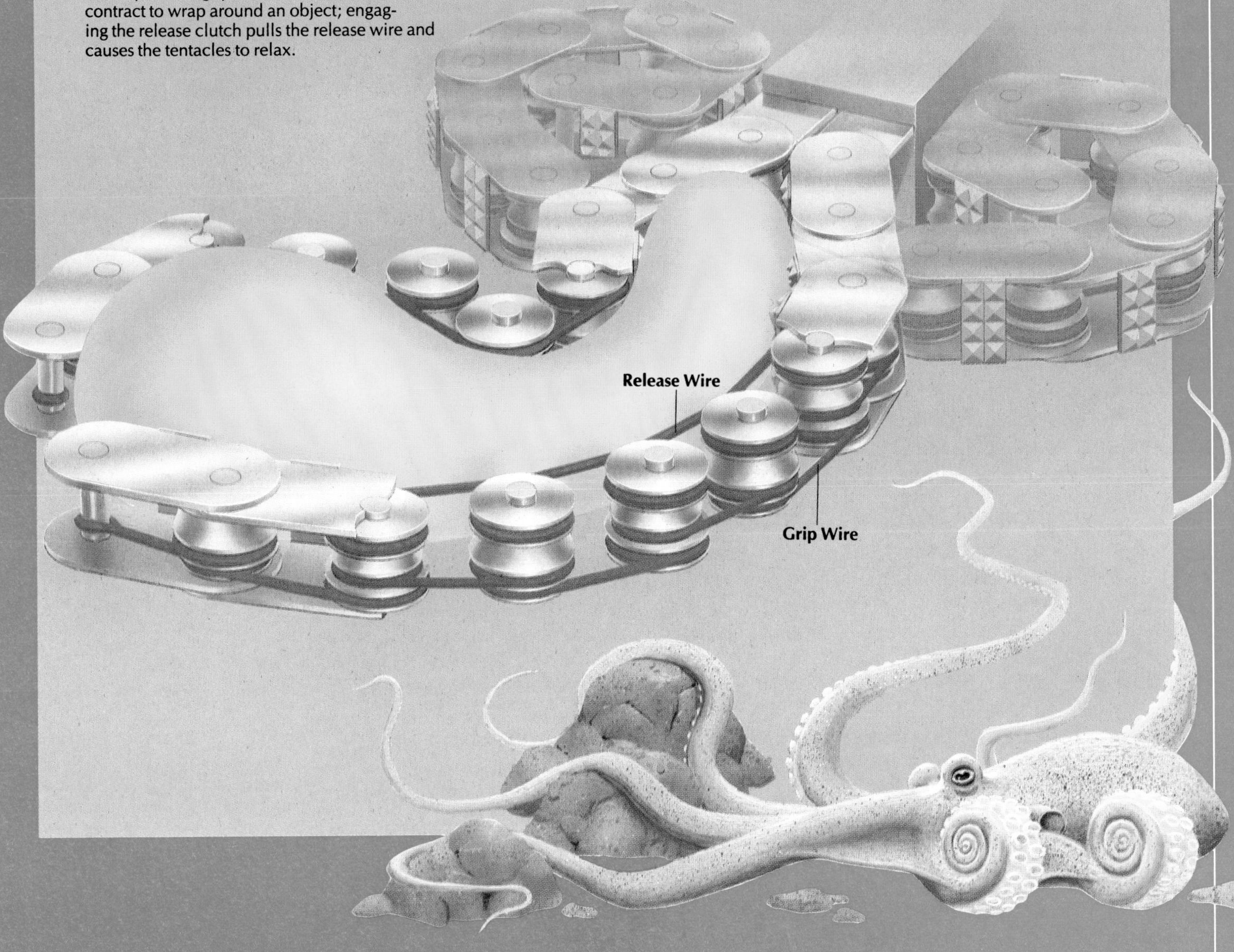

Release Wire

Grip Wire

biological organisms have evolved. "Many of the real-world constraints that shaped life," Moravec contends, "also affect the viability of robot characteristics." And Moravec is convinced that mobility is the key to truly intelligent robots. Intelligence as we think of it "has evolved exclusively in creatures that have a mobile way of life," he says, "and in the evolutionary record, they were mobile before their nervous systems grew." A clam—or, to take the extreme case, a tree—has little or no need of intelligence. Conditions do not change very much in the world of a clam or a tree. On the other hand, octopus and squid—mollusks that left their shells behind to roam—encountered a varying and unpredictable world and had to develop general intelligence to cope with it.

A NAVIGATING CARD TABLE

Moravec spent the years from 1973 through 1980 as a graduate student at Stanford University's Artificial Intelligence Laboratory, laboring the entire period over a "modest attempt," as he wrote in his doctoral dissertation, "at endowing a mild-mannered machine with a few of the attributes of higher animals." The machine, known as the Stanford Cart, had been languishing at the lab in various stages of incompleteness since 1966. It looked something like a tall card table with motor-driven wheels. Moravec improved its video-camera sensing system and rebuilt its radio link to a computer. He then programmed the cart to note the presence of obstacles in its path—chairs, boxes, cars and trees, for example—and to steer around them. The system, as with so many other mechanical pioneers, was excruciatingly slow. "The cart moves one meter every 10 to 15 minutes, in lurches," Moravec reported. "After rolling a meter, it stops, takes some pictures and thinks about them for a long time. Then it plans a new path, executes a little of it and pauses again." Negotiating a 20-meter-long indoor course took five hours, but at least the cart performed reliably. An outdoor run proved less successful. The harsh glare of the sun generated more contrast than the camera could handle. In addition, the cart identified shadows as obstacles and became confused because the shadows would shift while it was pondering its next move. It avoided two obstacles but collided with a third.

Nonetheless, the cart demonstrated that a robot could navigate in an unstructured environment. A Mars rover would represent one use for such a device, Moravec noted, anticipating Carnegie Mellon's Planetary Rover by some years. Any robot on Mars would have to operate largely on its own, because data would take half an hour to reach the earth and return, making it impossible for earthbound controllers to order a sudden course correction. Moravec viewed "more mundane applications," such as a robot servant or a self-steering car, as especially demanding because of complexity and cost constraints. "On the other hand," he wrote, "work similar to mine will eventually make them feasible."

When Moravec arrived at Carnegie Mellon's newly created Robotics Institute in 1980, he found a cadre of kindred spirits, roboticists who were fascinated by the problems of robot way-finding. Over the next few years, they produced a series of machines. One, Neptune, served as a test-bed for new sensing systems. Sturdy and small, built somewhat like a tricycle, it was limited mostly to indoor work because power was delivered by umbilical cord; moreover, its electrical motors ran at only a single speed. At times, Neptune navigated with the aid of a pair of video cameras that saw in stereo, like human eyes, to generate a three-dimensional image.

At other times, a ring of 24 sonar sensors was mounted atop Neptune. The sonar accurately measured distances to objects, but gave only rough indications of their location. To place obstacles more precisely, researchers programmed the robot to create a so-called certainty grid. The grid software divided Neptune's surroundings into squares measuring four inches on a side. The program then combined the first set of 24 sonar readings to determine how likely each square was to be obstructed. As the sonar continued to take new measurements, the probabilities were modified accordingly, until the grid stabilized into a map by which Neptune could steer. Certainty grids proved so valuable to robot navigation that they were also used in the programming of Uranus, a four-wheeled automaton that replaced Neptune in 1988.

Uranus and its relatives are strictly indoor robots, but Carnegie Mellon has also produced a series of outdoor research systems designed to handle bumpy terrain and shifting light. First of these was the Terregator. About the size of an office desk, powered by a gasoline generator and capable of driving its six wheels at different speeds, the Terregator gave the roboticists a capacious, versatile platform for their sensors and control devices. It operated outdoors on a so-called soft tether — a radio link on two UHF channels that transmits data to a computer. The vision system tried to keep the robot on a path or road by looking for abrupt shifts from light to dark, indicating road edges. It was often fooled, however, by shadows — the bugaboo that tripped up Moravec's Stanford Cart.

More recently, the Terregator has been used to test special-purpose robotic sensors, including magnetometers and ground-penetrating radar for surveying hazardous waste sites. Outdoor navigation research is now the province of the NavLab, a massive robot based on a modified Chevrolet van. Researchers drive NavLab to off-road test sites, then turn control over to built-in navigational computers. The passengers remain aboard, the better to observe how well laser range-finders and color video cameras enable the computers to steer NavLab across its testing grounds.

THE ROAD TO GENERAL COMPETENCE

As the presence of ride-along programmers suggests, self-guiding vehicles are likely to remain research instruments for a while. Moravec sees no reason to worry about applications for the moment. He views these vehicles as broadening robot capabilities in the most fundamental ways. "Solving the day-to-day problems of developing a mobile organism steers one in the direction of general intelligence, while working on the problems of a fixed entity is more likely to result in very specialized solutions. Mobile robotics may or may not be the fastest way to arrive at general human competence in machines, but I believe it is one of the surest roads."

Some scientists, however, have given detailed thought to real-world roles for self-guiding vehicles. For example, John McCarthy, a Stanford University professor of computer science who back in 1974 had supported Hans Moravec in his work on the Stanford Cart, thinks that a computer-driven car may become feasible in the not-too-distant future. It would need a powerful computer that today could cost anywhere from $400,000 to $800,000; but within a decade,

Refining Robot Skills

The challenge of equipping robots with the skills to operate independently, outside the regimented confines of a factory or laboratory, has taxed the ingenuity of academic, military and industrial scientists for years. The 11 experimental devices on these two pages represent some of the milestones of more than two decades of research into robotic navigation, walking and manual dexterity. The Tomovic hand, for example, designed by Rajko Tomovic at the University of Belgrade, was one of the earliest attempts to equip a machine with human-like tactile feedback. The manipulator built jointly by researchers at Stanford University and the Jet Propulsion Laboratory (JPL) carried those efforts further still. Carnegie Mellon's Terregator and the U.S. Defense Department's Autonomous Land Vehicle (ALV) are only two of several projects geared toward building machines that can navigate with the aid of a vision system or other sensors. Neither device can yet handle itself in unstructured surroundings without supervision. But all of the projects shown here may be succeeded by progeny more capable of interacting with and adapting to the unpredictable physical world.

Terregator. Carnegie Mellon's gasoline-powered terrestrial navigator has served as a test-bed for outdoor-navigation programs.

One-legged Hopper. The size of a two-year-old child, a balancing robot designed by Marc Raibert in 1983 can hop down pathways at 5 mph.

Tomovic hand. The pressure-sensitive fingertips of this 1965 aluminum model prefigured teleoperator devices equipped with sensory feedback.

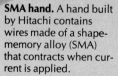

SMA hand. A hand built by Hitachi contains wires made of a shape-memory alloy (SMA) that contracts when current is applied.

Utah/M.I.T. hand. The four fingers of this 1985 design move at five times human speed and have the grip of a firm handshake.

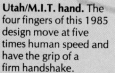

Stanford/JPL hand. Built in 1982, this three-fingered hand has sensors that detect both force and position.

ASV. Ohio State's three-ton, 17-foot-long Adaptive Suspension Vehicle is built to cover rough terrain on six hydraulically powered legs.

ALV. The Autonomous Land Vehicle is designed to navigate with the aid of sonar, TV cameras, inertial-guidance sensors and laser range finders.

Stanford Cart. Interpreting images of its surroundings, the cart could maneuver around objects at the rate of 10 to 15 minutes per meter.

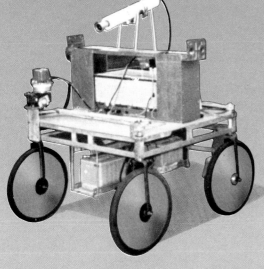

Shakey. In 1969, Stanford Research Institute's wobbly robot pioneer applied logic-based problem-solving to such tasks as negotiating a cluttered room.

Titan III. The Tokyo Institute of Technology's breadbox-size quadruped can mount stairs with the aid of sensors in its footpads.

that cost might be a tenth as much. McCarthy envisions the vehicle as responding to commands punched in on a keyboard: Go to a certain destination, stop at that restaurant, slow down, do not pass, return home. "The user need not be a driver," he has written. "This would permit children, old people and the blind greater personal freedom." The user, in fact, need not even accompany the car. You could drive to work, says McCarthy, and send the car home for your spouse's use; at the end of the workday the car could return on its own to pick you up.

In theory, self-guiding cars could greatly enhance automotive safety. But no one is proposing to fund the development of such a car, and McCarthy acknowledges glumly that auto makers and computer companies are unlikely to rush forward to do so.

IN FAVOR OF MANUAL SPECIALISTS

There is abundant interest, however, in devising dexterous robotic hands — yet another of the features that, like walking or navigation, would make possible truly resourceful and versatile robots. But what sort of hands? One school of thought favors specialization. Instead of creating multifingered, general-purpose hands of the human type, say the proponents of this view, researchers should focus on developing different hands — or, more accurately, end-effectors — for different tasks, with claw, gripper, electromagnet, hook or other specialized tool snapping in place as each task dictates.

Under tightly defined conditions — repetitive assembly-line functions, for example — the argument makes sense. If many special end-effectors are needed, however, the expense of maintaining a sizable inventory may be high. M.I.T. computer scientist Tomas Lozano-Perez offers a more important rebuttal — the point so often emphasized in research centers: "One of the primary advantages of robots is their generality and programmability." A robotic version of the human hand may well increase the possibility that robots can work at a multiplicity of chores in an unstructured environment — in space, on the sea bottom and perhaps, one day, in the home.

John M. Hollerbach, of McGill University in Montreal, Canada, is heartened by the advances on this particular frontier. "Already there has been more novel design in robot hands than in most other areas of robotics." Still, the challenge of designing hands to rival the human version is daunting. Both mechanically and as a sensing tool, the human hand is a marvelous creation. The tendons of the hand are lubricated by synovial fluid and encased in sheaths, producing vanishingly low friction. Its skin has four kinds of sensory nerve endings for touch alone, spaced as closely as one for every square millimeter at the fingertips. Each of the four types has a specific function. One serves for very fine discrimination, enabling a person to sense the curvature of a marble that is held between finger and thumb, for example. A second type of sensor responds only if the skin is moving — as, for instance, when fingertips are run over a rough surface. A third type detects vibrations, such as those produced by a tuning fork held in the hand or applied to the skin. And a fourth type can sense the stretching of skin, which might indicate when an object is about to slip.

Most designs for an all-purpose hand are at least vaguely anthropomorphic, with either three or four digits — two or three fingers plus an opposable thumb.

Three is the lowest number of digits needed for true dexterity: Two can hold an object while the third acts on it. A three-fingered hand has the added ability to grasp odd-shaped objects that a simple gripper, with its parallel jaws, cannot.

The first significant multifingered hand came from Japan in 1977, when a three-fingered hand was developed by Tokuji Okada of the Electrotechnical Labs in Tsukuba Science City, a research center 37 miles northeast of Tokyo. The hand could slowly twirl a baton, roll a ball in its fingers and turn a nut onto a bolt after threading it — a set of skills that has become a standard for dexterity. But it could not perform repeatedly and reliably. Its steel-wire tendons stretched, and the hand had no sensors to measure the pulling force of the tendons. The hand has since been retired to a museum.

The first major advance in the United States was the Stanford/JPL hand, developed in 1982. Ken Salisbury designed it for his doctoral dissertation at Stanford University, with Carl Ruoff collaborating at the California Institute of Technology's Jet Propulsion Laboratory. Salisbury did everything possible to hold complexity down. His hand used the minimum three digits and the fewest possible tendons, joints and sensors; it represented a research tool pared to the essentials. It could roll a soft-drink can in its fingers and also move the can in other ways at the same time. One of Salisbury's robot hands is at M.I.T., where he now works; ten other copies of the Stanford/JPL hand are at other research centers.

CABLES FOR TENDONS

The most promising of the current crop is the Utah/M.I.T. hand, a combined effort directed by Stephen C. Jacobsen of the University of Utah's Center for Engineering Design and McGill's John M. Hollerbach, who was at M.I.T. during the hand's early development. Its thumb is at the bottom of the palm rather than at the side. In early versions of the hand, tendons were polymer tapes that rolled across tiny pulleys at the direction of computer-driven air pistons. By 1990, the tapes had been replaced by polymer cables for durability.

The fingers of the Utah/M.I.T. hand can flex at about five times human speed and grip with comparable force — and delicacy. At one demonstration, the hand was put through its paces by remote control. It gently grasped an egg, cracked it and emptied the contents into a bowl. Then one finger, capable of whipping back and forth at 65 strokes per second, beat the egg.

To manipulate an object under human direction, no matter how cleverly, is not enough. A robot operating on its own would need hands that could sense an object's shape, orientation and, to prevent the object from slipping away, its degree of contact with the fingers. Vision can tell the robot that the object on the table is a bolt, and even describe its orientation, but it cannot provide feedback to let the robot know whether the bolt was actually picked up correctly. A tactile sense can. Moreover, a tactile sense can substitute for vision in tight or dark places.

One kind of sensor being developed for use on the Utah/M.I.T. hand is based on the fact that when the distance between two closely spaced electrical conductors changes even minutely, the electrical relationship between them, or capacitance, changes too. In the mid-1980s, M.I.T. graduate student David M.

Siegel incorporated this principle into a tactile sensor made from a rubber square half an inch on a side. Eight thin, parallel, conducting lines of copper were printed on the top of the square, and eight more, turned 90 degrees, were pressed underneath the rubber square. The intersections of the top and bottom conductors, though separated by the rubber, formed a matrix of 64 tiny capacitors. Any pressure on the rubber pushed the conductors closer together at that point and altered the capacitance. The change could easily be translated into information that told the hand how much pressure was being brought to bear, and it could even reveal the shape of the item exerting the pressure. Plans are to install the sensors on each finger of the Utah/M.I.T. hand.

Plenty of other sensor designs are competing for attention. One possibility is a kind of plastic film called a piezopolymer that, like the ceramic phonograph cartridges of the past, generates electricity when pressure is applied. Fiber-optic sensors are also being tested; these devices emit beams of light onto a reflecting surface and measure variations in the reflected light as the surface is moved by pressure. Yet another approach is a solid-state sensor — something like an electronic chip with the innards exposed and then covered with conductive rubber. Pressure on the rubber changes the electrical resistance on the surface of the chip and transmits an electrical description of the pressure. Solid-state sensors work well, but because such chips are built on a rigid plastic or ceramic base, the sensors lend themselves more to being mounted on a flat surface than on the curved or irregular surface of a robotic hand. Still other researchers have been considering arrays of strain gauges — sensors capable of detecting the direction of forces as well as their strength. Strain gauges could sense when an object in the hand was twisting or slipping, simply by measuring the changing directions of the force it exerted on the hand.

Plainly, robot hands — like robot legs or eyes or reasoning powers — have a long way to go before they can approach what biological evolution has achieved over the course of hundreds of millions of years. Much more will have to happen in laboratory hatcheries around the world before robots warrant comparison to nature's handiwork. Yet there can be little doubt that the journey toward self-reliant machines has begun.

Strategies of Legged Locomotion

3

The invention of the wheel has long been hailed as a milestone in human progress. But wheels do best on relatively smooth surfaces such as roads or railroad tracks; they are ineffective on ground that is soft or uneven. Tracked vehicles can manage uneven terrain, but not mountainous or swampy territory. In these situations, the advantages of legged locomotion become apparent: Animals and humans can choose the individual footholds that offer the best support; a wheel or track requires a path of support that is essentially continuous.

Generally, legs also have an advantage over wheels in and around buildings because they can climb stairs, step over obstacles and negotiate narrow spaces. One potential application of a legged machine, for example, might be to rescue people or property from burning buildings. Another could be to perform a variety of inspections in a nuclear reactor plant while the plant remained in operation.

Despite the abundance of models to be drawn from nature, roboticists working on legged locomotion face a considerable challenge, only partly alleviated by the advent of microelectronics. Since the 1980s, it has been possible to build computers into legged vehicles, but the problems of coordination, balance and negotiating rough terrain, while not insurmountable, have proved difficult to solve.

As illustrated by the imaginary robot on the following pages, several classes of gait, including creeping, walking and running — as well as such activities as climbing — are possible for legged creatures (whether biological or mechanical). In attempting to build machines capable of even one of these gaits, researchers have had to decide how to allocate the available computing power. Some, for example, have chosen to build machines with so-called static stability — that is, the robot is designed to move from one stable stance to another, and its control computers need only deal with equations relating to the coordination of the legs and to centering its mass over its supporting base.

By contrast, a one-legged hopper *(pages 69-70)* can afford to devote more attention to balance because the problem of coordination has been eliminated. In this case and in others, the emphasis is on active balance; the legged system must incorporate its speed and the kinetic energy of its own mass into its programming. Without active balance, a running machine with its center of gravity too near the front feet would fall over if it needed to stop abruptly.

Creeping: A Simple, Stable Gait

The simplest class of gait — and the one researchers understand best — is creeping, in which one foot at a time is off the ground. A quadruped robot would thus use a stable tripod as a base and move its fourth leg forward; it would then shift its weight to another combination of three legs, lifting the free leg and moving it forward. Six-legged insects, such as the beetles shown here, raise more than one leg at a time, but their strategy for maintaining stability is equally effective: alternating tripods with two legs on one side and one on the other.

The stability inherent in creeping makes it the easiest gait

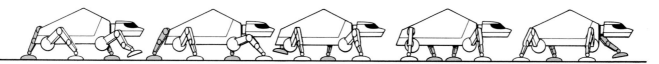

A legged robot might use the stable creeping gait to keep from falling over in a high wind or while crossing a shaky bridge, as shown below at right. Only one leg moves at a time, with the weight shifting slightly forward after the foot is planted. A cycle of the gait for a quadruped (above) begins with the right foreleg moving forward, followed by the left rear leg, then the left foreleg. To complete the cycle, the right rear leg will move forward next.

for a robot's computers to control, but the programming challenges are still considerable. Computers must monitor and direct the movement of all the joints in each leg and keep track of the position of the feet currently on the ground. They must also prevent the legs from banging into one another and avoid the splay-legged collapse that would occur if the robot's legs were moved too far in one direction. This complex job could be divided into three levels: At the servocontrol level, the movement of each joint is corrected according to information from sensors. At the coordination level, the legs are ma-

nipulated so as not to collide with one another. At the third level — which is yet to be fully achieved and might be called logic — the robot picks its foot positions and plans its route.

Research on legged machines must also take foot design into consideration. Animals have evolved feet that serve them best: broad feet for swampy ground or unstable conditions, small, hard feet for nimbleness in rocky terrain and so on. In time, legged robots, like the ones shown here and on the following pages, may be able to change feet to suit their needs just as industrial robots now change end-effectors (page 49).

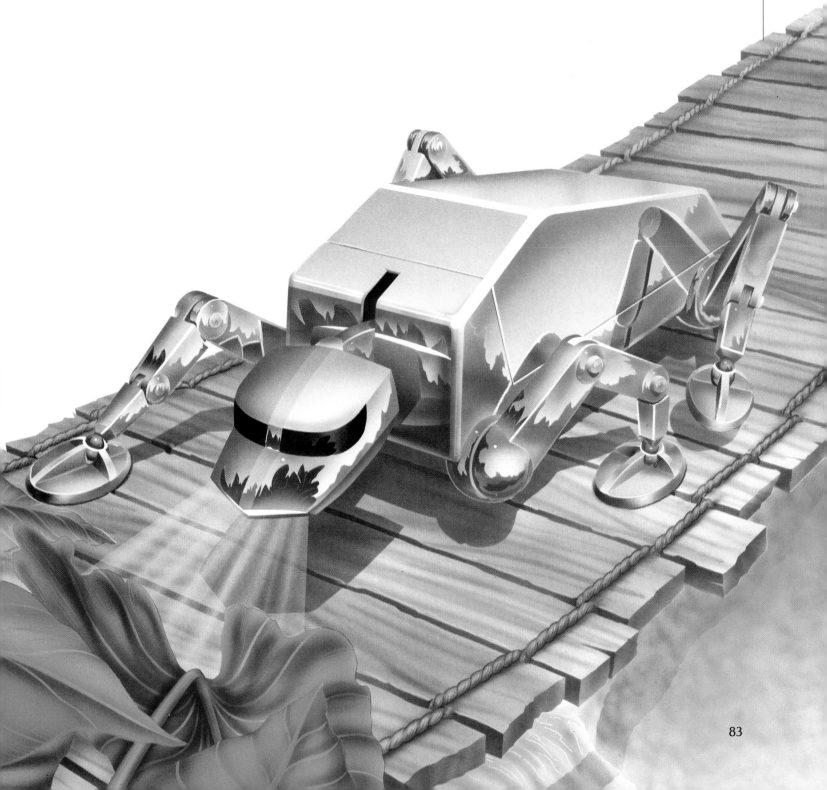

Walking: Good for the Long Haul

In studying a gait in which only one foot at a time is off the ground and moving forward, researchers limit the number of factors they have to plug into the computer's calculations. When a legged machine creeps, it moves so slowly — transferring its weight from one stable support pattern to another — that its forward momentum is minimal, and it can stop at any time without falling over.

Walking is a more complicated activity. Researchers who

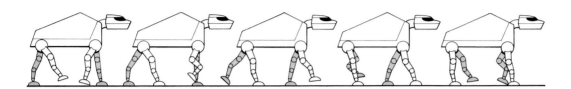

The fall-and-recover cycle (left) begins with the robot in an unstable state: Only its two left legs are on the ground. But it quickly plants the right rear foot to create a tripod. It then lifts the left rear foot, leaving its weight on a diagonal pair before falling to the right front. Finally, its weight shifts to the two right legs.

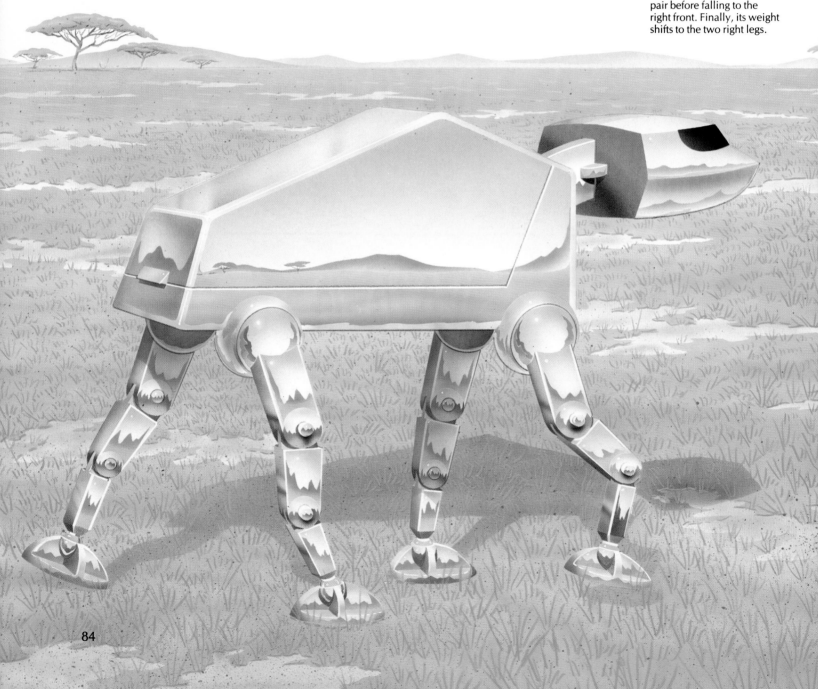

have studied human and animal walking have dissected the movement of the legs into a coordinated cycle called fall-and-recover. As demonstrated by the four-legged robot shown here, the walker lifts one leg from a stable three-legged position to pass briefly through a two-legged stance before falling into another stable tripod. For example, a horse typically employs repeated sequences of four different pairings of legs when walking — two lateral and two diagonal.

The complexities of such a gait clearly demand more computing power than is needed with creeping, but this is offset by greater speed. Walking is also more energy efficient than any other class of gait. The key to this efficiency lies in simple physics: When an animal moves at its optimum walking speed, each raised leg swings freely, like a pendulum, thus requiring little expenditure of energy to move from one position to another.

Running: The Problem of Balance

Running, the fastest class of gait for human or animal, nearly turns a legged creature into a flying machine. Frequently, only one foot is touching the ground, and sometimes — as with the cheetah below — all feet are momentarily airborne. Because much of the energy of each stride is spent launching the body into the air, running is far less efficient than walking or creeping; in the animal kingdom, it is generally reserved for short spurts of flight or pursuit.

Active Balancing

One researcher's approach to running breaks the problem of balance and propulsion into three parts: Controlling forward speed, controlling rhythmic up-and-down movement and controlling the tilt — the roll, pitch and yaw — of the body. The robot's computer integrates the results of these computations into a single set of instructions for each leg; sometimes only one leg is in contact with the ground to provide the calculated push that keeps the robot steady and on course.

For a robot, running would require more sophisticated computer control than the other gaits because when contact with the ground is intermittent, balance becomes critical. In one experimental approach — applied to a one-legged hopping robot — the machine uses an inertial navigation system that constantly senses, among other things, the attitude, or tilt, of the robot with respect to its rapidly changing surroundings. The robot's computers process this data and initiate adjustments to the machine's joints to keep it from falling over. In the real world outside a research lab, a robot's computers would also need to take into account factors that affect the way the feet push off, such as the texture and resilience of the ground. And because the running machine moves so fast, more computer power must also be applied to the problem of detecting and identifying obstacles in its path quickly enough for the machine to avoid them.

In a running sequence based on a horse's canter, the robot pushes off on its left rear foot while its other legs are airborne. It balances briefly on its right rear foot before the left forefoot touches down. The right forefoot then bears the weight for the next phase.

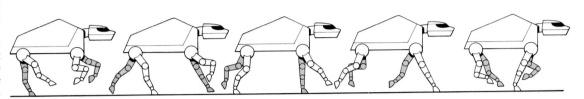

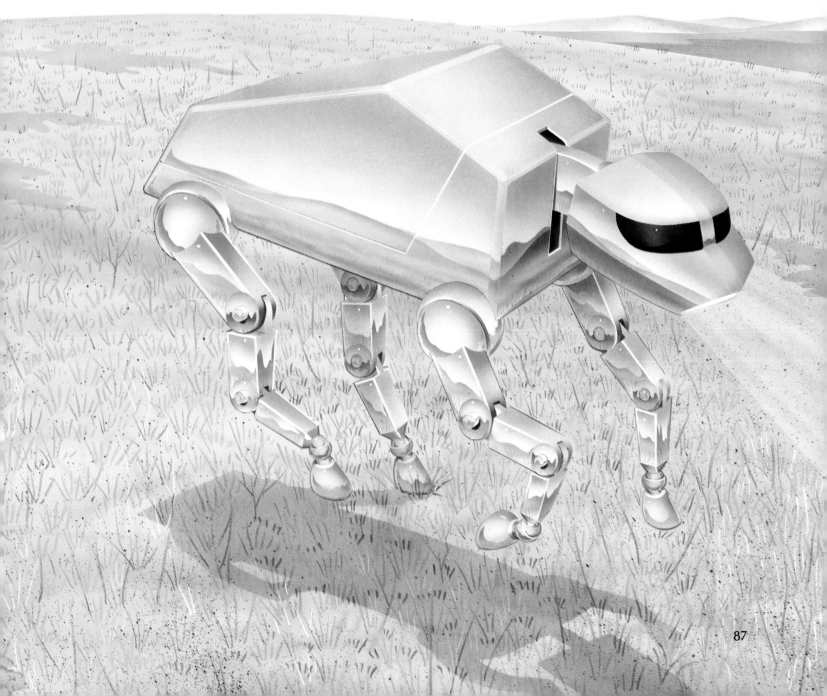

Climbing: A Hunt for Sure Footing

On rough terrain where footing is uncertain, the robot cannot rely on the regular sequences that characterize creeping, walking and running. Instead, it must use the patternless motion of climbing — extending its legs one at a time to probe for secure footing. Climbing puts much heavier demands on the computer control system than do the sequenced gaits, because each foot placement must be considered as a separate case — taking into account not only the robot's balance and coordination, but also the question of which leg is the best one to move next and whether the surface is able to support the machine's weight.

One way to reduce the computing demands of climbing is with a technique called follow-the-leader. The front legs are used to find solid footholds; the rear legs are moved only to the spots already proved safe by the front legs. The gain in simplicity comes at the price of a loss of speed and flexibility; follow-the-leader is probably not the most efficient way to traverse rough terrain quickly.

Climbing also requires that the robot be able to plan its route; otherwise, it might pick its way, step by step, to a point where progress is blocked by an insurmountable obstacle. Extremely sophisticated artificial intelligence programs will be needed to give such a robot the route-planning ability of a climber as experienced as the mountain goat.

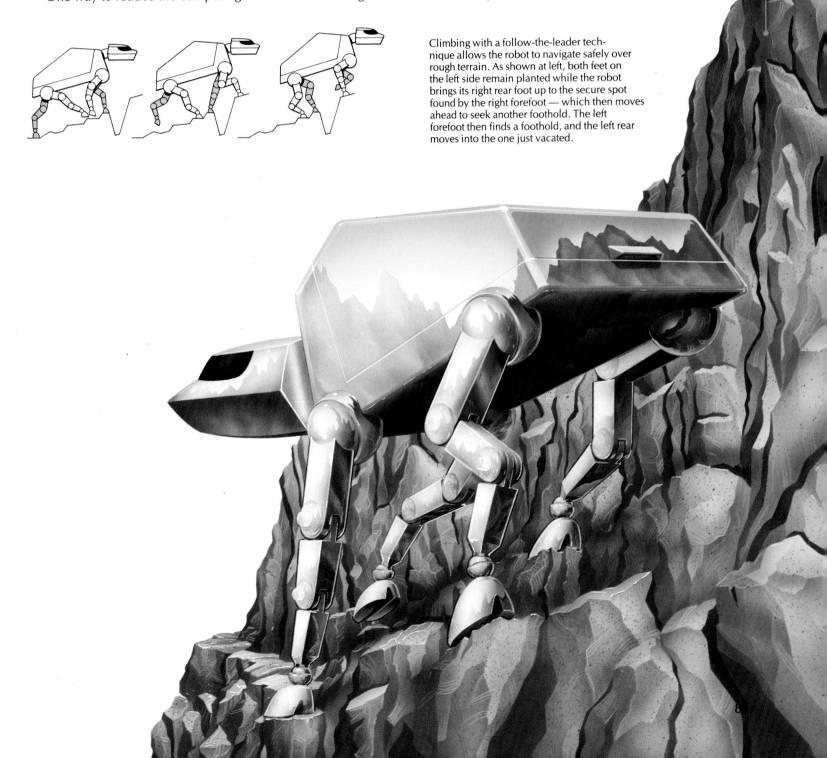

Climbing with a follow-the-leader technique allows the robot to navigate safely over rough terrain. As shown at left, both feet on the left side remain planted while the robot brings its right rear foot up to the secure spot found by the right forefoot — which then moves ahead to seek another foothold. The left forefoot then finds a foothold, and the left rear moves into the one just vacated.

Duplicating
Dexterity
and Touch

The human hand is a remarkable instrument. Despite the complexity of its parts, it has an economy of structure (namely, a thumb that can be made to oppose each finger) perfectly adapted to the hand's essential functions of grasping and manipulation. Equally important to human evolution, the hand is also an extremely accurate sensory organ, capable of gaining information about objects merely by touching them. In this way, the hand provides the brain with data unobtainable through vision, such as the temperature, texture and weight of objects. Hand and brain thus form an inseparable combination that allows humans to manipulate the world at large.

Human-like hands for robots are desirable because of the wide range of tasks they would make possible. In industry, a dexterous, sensitive machine hand could put together complex assemblies or inspect welds on the inside of objects. Such a hand would greatly increase the ability of remotely operated robots to do such disparate jobs as undersea salvage, agricultural harvesting and spacecraft repair. And dexterous hands would certainly be an important characteristic in any design of autonomous, intelligent robots in the distant future.

The shape of the hand is a significant consideration for dexterity; the juxtaposition of rounded and flat surfaces on the fingers and the palm of the hand determines how well the hand can manipulate objects of different sizes and shapes. By incorporating different combinations of surfaces into a hand, designers can increase its versatility.

In determining the overall size of the hand, roboticists must consider the situation in which it will be used. Generally speaking, if the robot is to work in surroundings structured for humans, the best size for its hand would be that of an average human hand. For manipulating very large or very small objects, however, or for working in constrained spaces, the average human-size hand might be inappropriate, and more suitable robotic adaptations would be necessary.

The artificial skin covering the hand plays a role in gripping and manipulating objects and can also serve as a sensory device that measures some of the characteristics of those objects. Patterns of ridges in the skin, for example, improve the grip and also provide a mechanism for sensing movement. In some cases, however, there is a conflict between these functions; for instance, soft skin provides a better grip than hard skin, but may inhibit the sensing of external phenomena.

Just as it does for humans, a general-purpose hand could supplement a robot's vision and help it learn about the world through the sense of touch. Because the hand's sensors are in direct contact with objects, tactile sensing avoids many of the problems inherent in vision, such as blocked lines of sight, bad lighting and the confusing clutter that makes it difficult to distinguish an object from its background. The sensors must therefore be built to withstand the wear and abrasion inherent in their job as well as the extreme conditions under which robots are often required to work. Attached to an underwater robot, for example, the sensors would have to be resistant to sea salt and cold temperatures. On a machine designed to clean up a nuclear power plant accident, the sensors would have to be able to function in spite of radioactivity.

The robotic hand shown on the following pages is a pure fiction, based upon concepts that have emerged from research into general-purpose hands and tactile sensing. Numerous mechanical constraints must be overcome before it is a reality. As yet, it is difficult to fit into a hand-size device all the mechanisms required to accurately control three joints on each of four digits; the same space would also have to accommodate the wires or fibers connecting the sensors to their power supply and to the robot's computer. And the sensors themselves, currently large, flat blocks of electronic equipment, will have to undergo considerable miniaturization.

But an even greater hurdle must be surmounted to bring a general-purpose robot hand into being: The robot to which the hand is attached must be provided with a high degree of artificial intelligence in order to manage it. To identify an object by its tactile characteristics, the robot must have reasoning ability and knowledge about the world; to develop and follow a course of action based on that identification, the robot must be able to learn and plan. Artificial intelligence programs with this level of sophistication have yet to be produced. Equally critical, such programs will require very fast computers to provide the real-time responses needed to deal with sudden events such as a bottle slipping from the hand's grasp.

A Tool for Exploration

With a general-purpose hand, an intelligent robot would be capable of discovering the characteristics of an object by manipulating it. As shown here, the robot might gather data through such exploratory strategies as stroking, squeezing and grasping. This information, once processed by the robot's computer, could provide the basis for deciding what to do with the object. In performing a task, the hand could then employ any of a variety of grasps to manipulate the object.

In a typical pattern of exploration, the robot would first try to determine the position and shape of the object by vision.

Then the robot would move its hand toward the object slowly, while its sensors watched for potential dangers such as excessive heat. The first touch would probably be very light, allowing other sensors to report data such as whether the object had sharp edges. The hand might then move across the surface of the object, sensing and reporting data about its shape, texture and thermal properties; the hand could gather additional information by grasping the object and manipulating it.

All of this sensory data would be continuously assimilated as the robot worked on, or with, the object. Depending on the size, shape and weight of the object, and on the job to be done, a robotic hand could choose from an almost infinite variety of grasps. For example, a power grasp (bottom right) would be called for in handling a heavy tool such as a hammer; a precision grasp (right) would be for a lighter tool such as a jeweler's screwdriver.

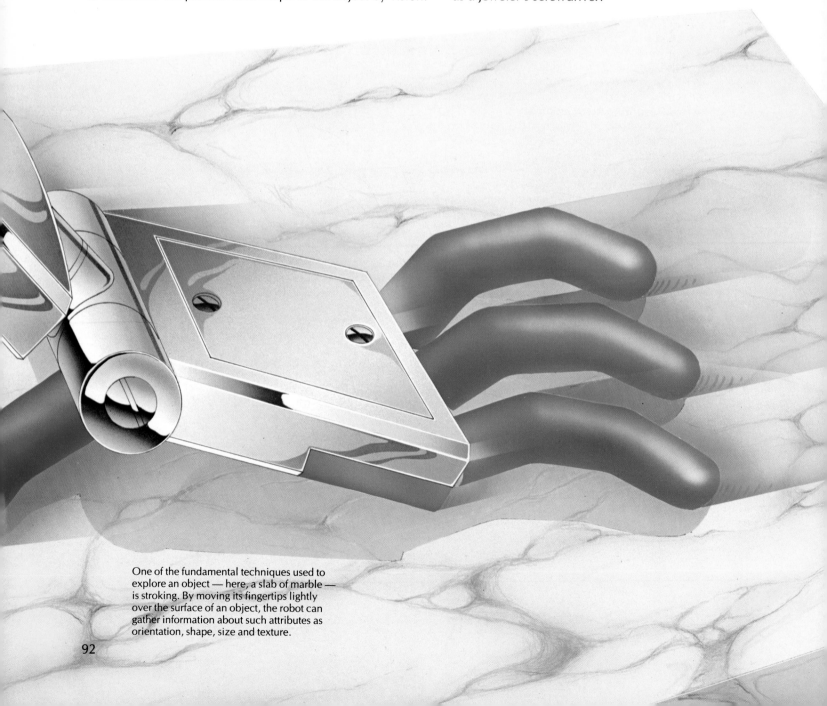

One of the fundamental techniques used to explore an object — here, a slab of marble — is stroking. By moving its fingertips lightly over the surface of an object, the robot can gather information about such attributes as orientation, shape, size and texture.

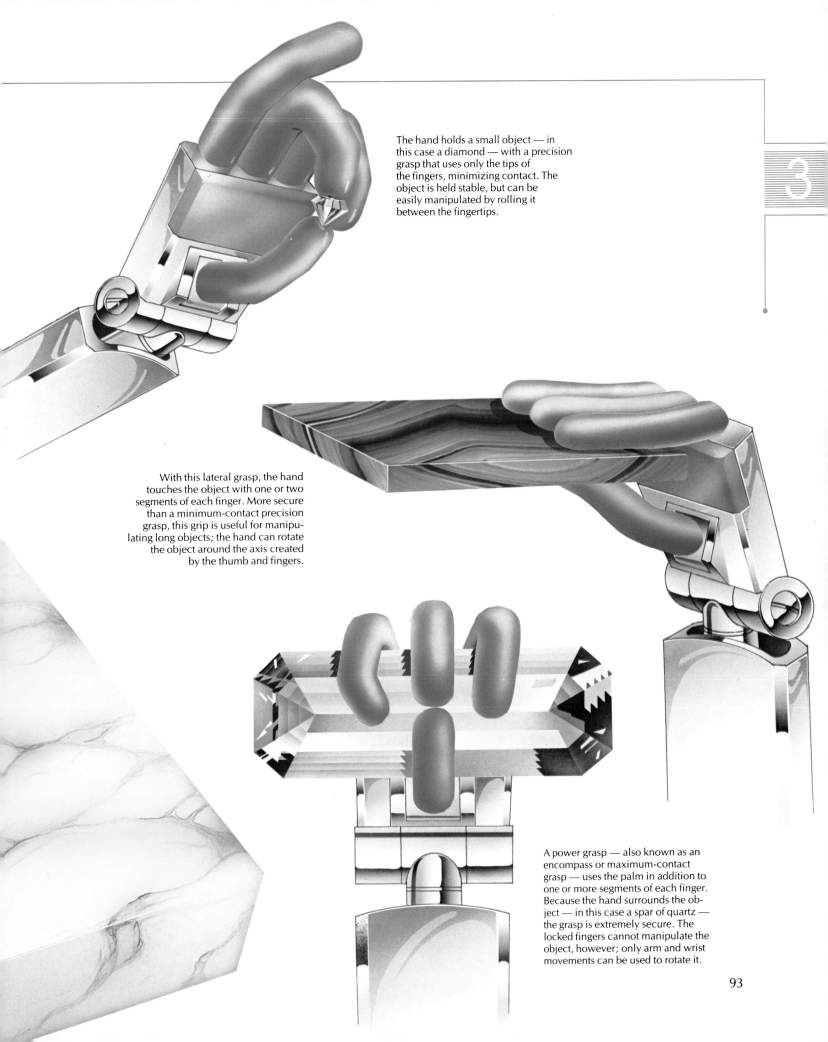

The hand holds a small object — in this case a diamond — with a precision grasp that uses only the tips of the fingers, minimizing contact. The object is held stable, but can be easily manipulated by rolling it between the fingertips.

With this lateral grasp, the hand touches the object with one or two segments of each finger. More secure than a minimum-contact precision grasp, this grip is useful for manipulating long objects; the hand can rotate the object around the axis created by the thumb and fingers.

A power grasp — also known as an encompass or maximum-contact grasp — uses the palm in addition to one or more segments of each finger. Because the hand surrounds the object — in this case a spar of quartz — the grasp is extremely secure. The locked fingers cannot manipulate the object, however; only arm and wrist movements can be used to rotate it.

93

Weight and Distinct Features

A robot with sufficient artificial intelligence could gain clues to the nature of an object by determining its weight and shape. These attributes would help the robot's computer identify the object as well as devise a strategy for handling it. A human-like hand could augment an industrial robot's visual ability in such complex tasks as exploring and sorting the contents of a bin filled with different kinds of machine parts.

In humans, the weight and general shape of an object are

A robotic hand equipped with sensors in the skin covering the fingers and palm could detect various points of contact with a grasped object — here, a baseball. Other sensors could record the angle of the wrist and of each finger joint. From this information, the robot's computer could infer that the hand held a sphere of a particular size. Meanwhile, force sensors at the wrist and finger joints could provide data for computing its weight.

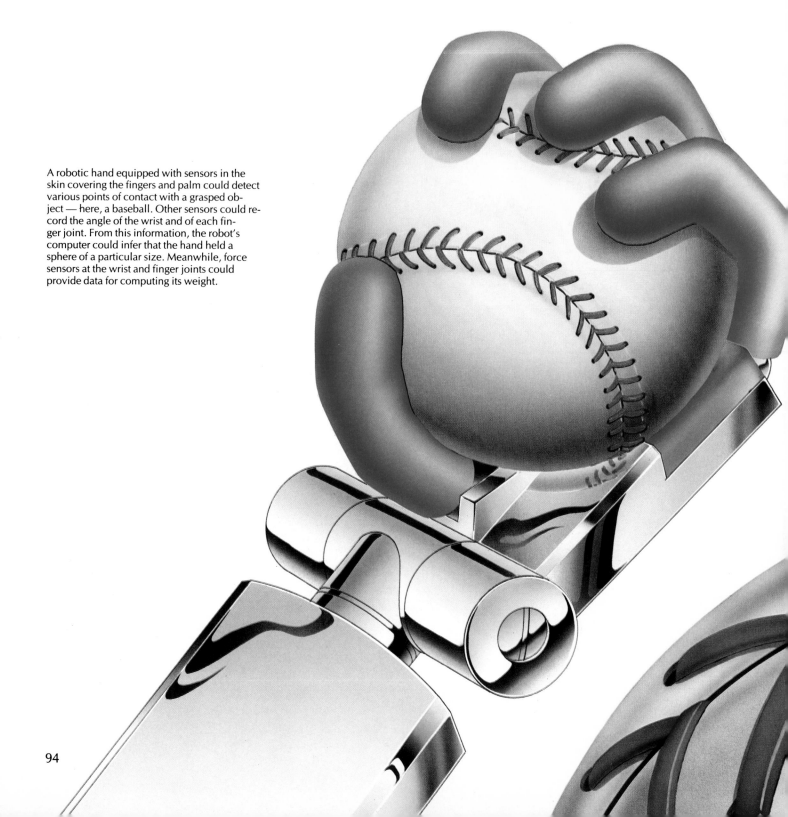

detected by a sensory technique known as kinesthetic sensing; in robots, this would entail the use of sensors at the joints of the fingers, wrist and arm to measure the angles of the joints and the forces brought to bear on them by the mass of a grasped object. Data on the angles of the joints would then be combined with information on points of contact between the object and the hand to allow the robot's computer to infer the overall geometry of the object. By analyzing the forces on the joints, the computer could calculate the weight of the object.

Other attributes that can be determined by touch are detected by so-called cutaneous sensing. In this activity, sensors in the robotic hand's artificial skin convey data about contact, texture, slippage and temperature to the robot's computer. Cutaneous sensing is used in shape detection to distinguish localized features that are too small to be sensed kinesthetically, such as bumps and holes on a surface.

In this close-up view, the compliant tip of one finger is compressed *(white bar, below)* by its contact with the stitches on the surface of the sphere. Sensors could detect the pattern of pressure distribution within the skin; given adequate programming, the robot could use this information to identify the sphere it holds as a baseball with stitched seams.

Sensing Slippage and Texture

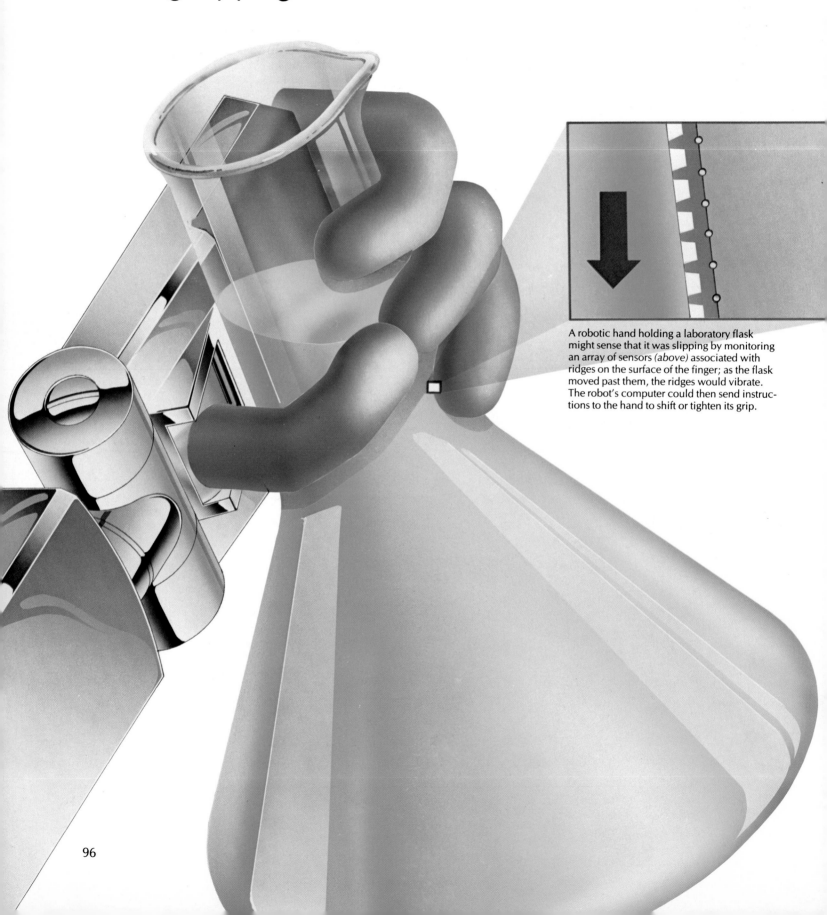

A robotic hand holding a laboratory flask might sense that it was slipping by monitoring an array of sensors *(above)* associated with ridges on the surface of the finger; as the flask moved past them, the ridges would vibrate. The robot's computer could then send instructions to the hand to shift or tighten its grip.

Two important kinds of tactile sensing involve motion. By moving the hand relative to the object, the robot could sense the fine features that make up texture. Conversely, by detecting the movement of an object relative to the stationary hand, a robot could determine that the object was slipping out of its grasp. Both types of sensing would require that the robot be able to detect vibrations of tiny ridges in the skin; in one scheme, this might be handled by sensors associated with each ridge (left).

The ability to detect slippage would be useful in industrial robots — provided, of course, that the robot could sense slippage in time to do something about it. So-called force sensors, which detect minute changes in the forces of contact between the object and the hand, would alert the robot's computer to the condition known as incipient slip, or microstrain. The computer, in turn, would have to be fast enough to process this information instantaneously, so that it could instruct the hand to adjust its grip.

Generally, a hand cannot detect the texture of an object if it is simply holding the object stationary. However, when the fingers lightly stroke a surface (below), the finger ridges brush against such minute features as hairline cracks, triggering vibration sensors. A robot's computer could interpret the resulting patterns of vibrations to determine the texture of the surface. This information could be used to identify an object in the absence of visual cues — in the dark, for instance, or deep inside a cluttered box — or for tasks such as distinguishing different grades of leather. In an industrial setting, the ability to sense texture would also allow a robotic hand to locate flaws and irregularities in materials such as glass or metal.

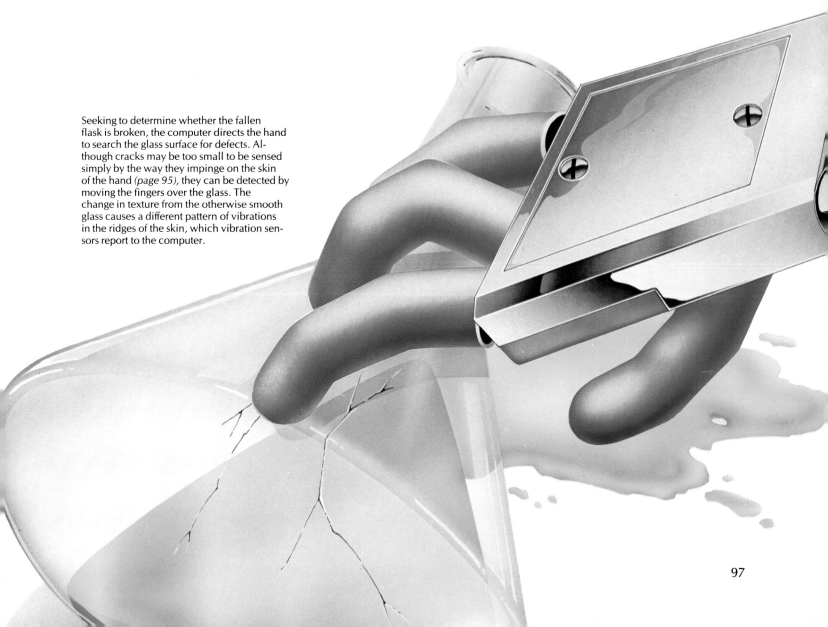

Seeking to determine whether the fallen flask is broken, the computer directs the hand to search the glass surface for defects. Although cracks may be too small to be sensed simply by the way they impinge on the skin of the hand (page 95), they can be detected by moving the fingers over the glass. The change in texture from the otherwise smooth glass causes a different pattern of vibrations in the ridges of the skin, which vibration sensors report to the computer.

The Dual Capacities of Thermal Sensors

The ability to detect temperature is an important tactile sense that does not necessarily require touching an object. To protect itself against dangerous and potentially destructive heat, the robot could be programmed to approach an unknown object cautiously while thermal sensors in the skin of the hand measured the heat radiating from the object. The robot's movement would have to be deliberate enough to allow its computer to process the sensory data and, if the object proved too hot, to withdraw the hand before it could be damaged.

A robotic hand equipped with thermal sensors could avoid damage by approaching a candle flame slowly. Such sensors would continuously report their readings to the robot's computer; if they detected a temperature too high for safety, the computer would instruct the hand to stop or withdraw.

A different temperature-related sense, one that does require the hand to be in contact with the object, is the ability to detect the flow of heat from hand to object. (This is similar to the human sense that perceives a piece of metal as colder than a piece of wood, even though both are at room temperature.) Because heat flow does not take place between objects that are at the same temperature, a robotic hand would have to be kept at a different temperature from that of its environment.

Measuring an object's thermal conductivity — the rate at which it absorbs heat — would thus give the robot another tool for distinguishing between different substances. For example, metal is highly conductive, absorbing heat from the hand, while cloth is a poor conductor, absorbing little heat. By combining this information with other data perceived by the tactile senses, a robot programmed with an appropriate store of knowledge might be able to draw fairly reliable conclusions about the identity of objects it encounters.

By measuring thermal conductivity — the rate of heat flow between the hand and a grasped object — the hand could distinguish between objects of identical shape and texture. Different substances may exhibit a wide range of flow rates, as shown below.

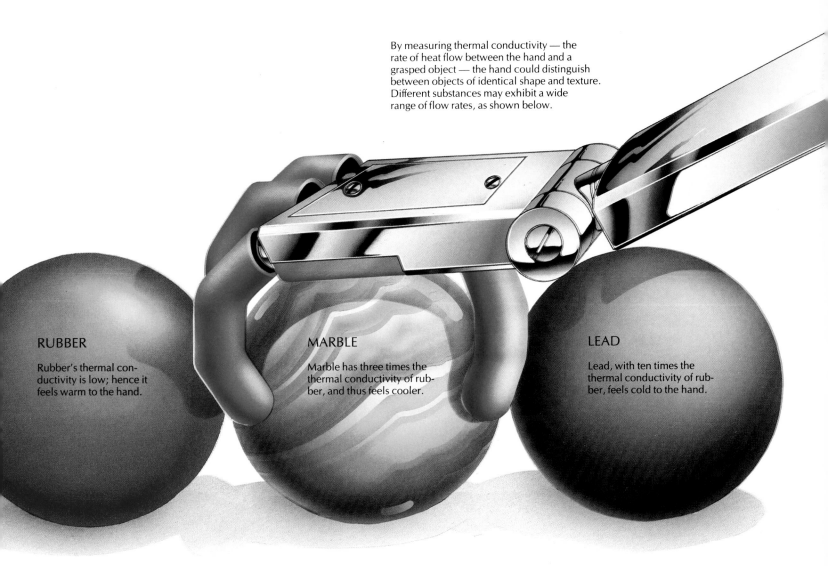

RUBBER

Rubber's thermal conductivity is low; hence it feels warm to the hand.

MARBLE

Marble has three times the thermal conductivity of rubber, and thus feels cooler.

LEAD

Lead, with ten times the thermal conductivity of rubber, feels cold to the hand.

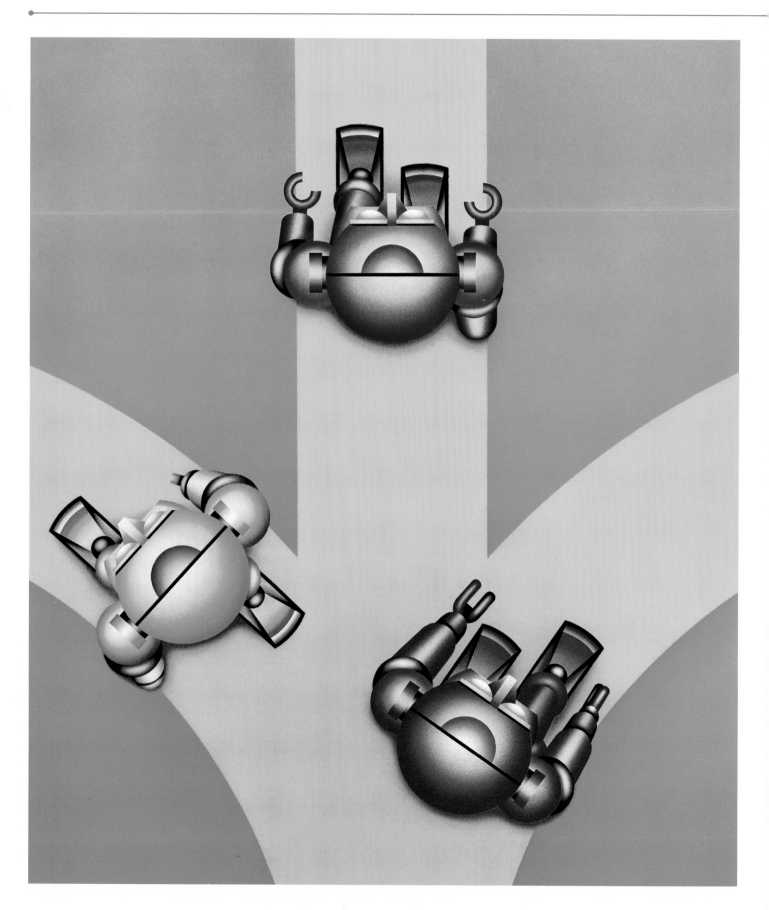

A Range of Superskills

Early in the morning of March 28, 1979, workers at Pennsylvania's Three Mile Island (TMI) nuclear power plant were trying to repair a minor malfunction in the water purification system of the Unit 2 reactor's secondary cooling loop. In so-called pressurized-water reactors like TMI's, three separate water systems, or loops, play a part in cooling the reactor's heat-producing core and converting that heat to electricity. During the dawn repair operation, pumps that supplied water to the secondary loop, which transferred heat from the primary loop to a steam generator, inexplicably shut down.

Because of various safety features built into the reactor, this development should have had no great consequences. But an auxiliary cooling system malfunctioned, and a calamitous chain of events was set in motion. Water in the primary loop, which picked up heat from the reactor core, was now unable to get rid of it at the boundary with the secondary loop; temperature and pressure in the primary loop began a steady buildup. Triggered by the pressure, a relief valve popped open. When this proved insufficient to reverse the situation, the reactor automatically shut down. As pressure began to drop, the relief valve should have closed. Instead, it stuck in the open position. The continuing loss of pressure caused the superheated water in the primary system to boil at the core. Pressure in the primary loop dropped so dramatically that the emergency cooling system designed to keep the reactor core from overheating was automatically activated. Infusions of cooling water flooded the reactor. But with the valve still open, there was not enough pressure to keep the water from boiling off.

In the control room, malfunctioning instruments led operators to believe that the reactor's core was safely awash in coolant, and they reduced the flow in the emergency cooling system. Pressure in the primary loop dropped again. With no water to cool them, the reactor's fuel rods overheated, their temperatures climbing to 2,500° F. — within 500° of meltdown. Radiation levels in the reactor containment building reached as high as 30,000 rems per hour, enough to kill a human being instantly.

Meanwhile, thousands of gallons of contaminated water had been pouring out the relief valve, overflowing a small holding tank and forming a seven-foot-deep radioactive lake in the basement of the containment building. A sump pump started drawing this water off to the adjacent auxiliary building, which had not been designed or intended to contain high-level radiation. Radioactive steam from both buildings started escaping through vent stacks into the atmosphere.

The initial series of events, unseen and unknown until hours later, took only minutes, but radioactive steam and gas continued to seep into the atmosphere for several days. Years later, the repercussions of the TMI breakdown, compounded by the much more serious accident at the Soviet Union's Chernobyl nuclear plant, were still being felt by the nuclear-power-generating industry.

Initial assessment of the damage at TMI, and even some of the cleanup, was carried out by human workers wearing many layers of protective clothing and operating for only brief periods inside the containment building. But the job of tackling the most seriously contaminated areas — the basement of the containment building and the reactor core itself — was clearly an assignment for creatures impervious to the effects of radiation.

Because robotics technology is still in its infancy, full autonomy for machines remains a distant goal, but a serviceable compromise has been around since the 1940s. Known as teleoperators *(pages 115-121)*, they are linked — often by a ribbon of cable — to human controllers who remain at a special console, safely away from the dangerous environment. The connecting link serves both as a tether and as a conduit for control signals, power and video data.

In a world not always hospitable to humankind, teleoperators and other robots with somewhat more autonomy can greatly extend the range of human powers and senses. They can serve as proxies at such accident sites as collapsed mines, fires and radioactive power plants. They can also perform certain kinds of tasks in factories or many fathoms below the sea. Some machines can navigate according to information from acoustic and other sensors, or detect infrared light or minute gradations of warmth indiscernible to the most highly sensitive person. When so designed, these robots may act as sophisticated fire alarms or seek out targets in long-distance wars. The programming of such devices — both the lifesaving and the lethal — is among the most complex in any area of computing.

At TMI, the first teleoperators brought in to act as cleanup crew were relatively simple devices, with limited skills. Over time, they grew increasingly capable. The initial machine, which arrived in August 1982, was a tiny, 25-pound tanklike vehicle called SISI (for System In-Service Inspection). Equipped with cameras and sensors, SISI was used to take photographs and radioactive readings of areas around the Unit 2 auxiliary building. The following spring, a six-wheeled, one-armed vehicle named Fred was fitted with a high-pressure water spray for decontaminating the walls and floor of a pump cubicle in the basement of the auxiliary building. Outweighing SISI by 375 pounds, Fred was able to lift 150 pounds with its arm and could extend itself to a height of six feet.

In November 1984, a 1,000-pound, six-wheeled remote reconnaissance vehicle (RRV) named Rover 1, one of a pair designed and built at Carnegie Mellon's Robotics Institute, made the first closeup inspection of the contaminated basement. The machine was lowered through a hatch from the floor above, and its three television cameras provided remote operators with a view of the damage, while two radiation sensors transmitted data that would help officials plan the next step in the decontamination of the site. More than four and a half years after the accident, radiation levels inside the basement topped 30 to 40 rems per hour, thousands of times higher than advisable for human safety.

But no one yet knew how much radiation had permeated the walls of the containment building. One year later, a Rover entered the basement again, this time equipped with a 1,000-pound core sampler. Its task was to drill into the thick concrete walls and obtain cores one inch in diameter, which scientists on the outside could test for radioactivity.

Guided by an operator at the remote-control panel, Rover's diamond-tipped drill module bit into the walls at a number of locations and withdrew a series of samples. Later analysis showed that the level of radiation lingering in the walls was still too high to permit a human work crew into the basement.

Once the full extent of the basement contamination was known, a massive robotic cleanup got underway. Laden with all manner of industrial cleaning tools, the Rovers gradually removed every form of radioactive debris capable of contaminating the building's upper floors. They flushed contaminated dust off overhead surfaces, removed as much as a quarter-inch of concrete from the walls, drilled holes into the walls through which they poured gallons of cleansing water, and pumped out 700 cubic feet of radioactive mud. Perhaps even more impressive than the robots' versatility was their extraordinary endurance. By the end of the project in May 1989 — more than ten years after the accident itself took place — the Rover team had served for almost five years without a failure.

TO THE BOTTOM OF THE SEA

The Rovers used at Three Mile Island are superb examples of robots venturing where humans cannot go. Similar remotely operated vehicles (ROVs) are serving as human surrogates in the inhospitable environment of the ocean deeps. Examples of underwater ROVs range from "flying eyeballs," which are essentially self-propelled underwater cameras, to more complex machines equipped not only with cameras but with sonar and manipulator arms.

Perhaps the most publicized contribution of the seagoing ROVs has been the exploration of the *Titanic* by an egg-shaped, lawn-mower-size flying eyeball called Jason Jr. — J.J. for short. (J.J. and a larger ROV, Jason, were named for the hero of Greek legend who recovered the Golden Fleece. The names are a rueful joke, alluding to Senator Proxmire's notorious Golden Fleece awards.) Designed and built by scientists at Woods Hole Oceanographic Institution in Massachusetts, the teleoperated J.J. carries still and video cameras capable of surveying 170 degrees side to side. What the robot sees, its operators also see, via a 250-foot tether that connects J.J. to the *Alvin*, a 25-foot minisub capable of diving to depths of 13,000 feet.

On July 13, 1986, expedition leader Robert D. Ballard and two others took *Alvin* 12,500 feet down to the *Titanic*, which a French-American mission headed by Ballard had located almost a year earlier, to begin 11 days of exploration. It was the first attempt to enter the "unsinkable" ship since an iceberg sent her to the bottom on her maiden voyage in 1912. Ballard and his crew first landed *Alvin* on the deck of the great ship, next to a gaping hole over the grand staircase that once had been covered by a huge, decorated glass dome. Ballard then released J.J. The tiny robot glided down the stairway — "down through four decks, like through a layer cake," Ballard recalled later — sending back images of everything it encountered. There was nothing left of the staircase itself; it had been devoured by wood-eating organisms. But an exquisite glass and crystal chandelier appeared out of the darkness, swaying in the underwater currents, as well as three- and four-foot-long stalactites of rust in vivid red, orange and yellow. Over the next few days, J.J. found chamber pots, corked champagne bottles and a shiny-handled safe with an orange crest. Directed to follow the mast up to the crow's nest, the machine discovered an intact brass mast light. Given other instructions, it slipped through the first-class entranceway to the ship's gymnasium.

In all, J.J. made four three- to four-hour explorations of the *Titanic*, passing its first deep-dive tests with flying colors. J.J.'s great success as an underwater explorer heralded more complex missions for Jason, which possesses two pincer arms, the ability to lift 20 pounds of samples from the ocean floor and a twin-video-camera system for stereo vision. Jason has since been teamed with a tethered underwater vehicle called Medea (another Golden Fleece allusion; this was the name of Jason's wife), which can carry sonar as well as video equipment. Medea affords its operators a panoramic view of the ocean floor while Jason swims out to perform the more specialized tasks of close-up inspection, making temperature probes and taking samples.

Perhaps the hardest-working ROVs are those associated with offshore oil rigs, undersea pipelines, and telephone and power cables. All offshore structures in the North Sea, for example, must undergo regular visual inspection by underwater eyeballs like J.J. And technological advances in the design of both manipulator arms and remote-control systems have led to an increasing number of ROVs like Jason that can do more than simply inspect. ROVs fitted with grippers or saws can help to lay or repair cables or pipelines at depths as great as 7,000 feet, where, as one industry spokesman has noted, "a 16-inch-diameter pipe with a one-inch-thick wall can collapse like a drinking straw flattened between your fingers" — as can the lungs of human divers.

One scenario calls for a team of ROVs, all controlled by tether. If a break in a gas pipeline is reported, technicians would turn off the flow of gas and dispatch a video-equipped survey ROV to look for the damage. When the mechanical scout found the spot, it would transmit the location through its tether to the mother ship and then guide a second ROV carrying repair modules to the site. A third ROV, called a strongback, would sever the damaged pipe with its hydraulically operated cutting arm and discard the damaged section before returning to the surface for replacement pipe.

The replacement pipe could include four flexible joints, two in the center and one near each end, to allow the strongback to bend the pipe into a shallow V-shape for easier insertion into the ends of the severed pipeline. Then the robot's hydraulic arms would push against the pipe, straightening it and locking the joints for a permanent, leakproof seal.

EVOLVING TOWARD AUTONOMY

Despite their capabilities, tethered ROVs are restricted by their need for a connecting cord to supply power and transmit communications. The farther a ROV's assignment takes it from the mother ship and the deeper it is required to dive, the more drag is created by an ever-longer cable. This causes difficulty in maneuvering and raises the ROV's power needs. Perhaps more serious, a number of ROVs have been lost when the tether was severed by the mother ship's propeller or when it got caught on some underwater obstacle.

One way to avoid this problem — and to create a nearly autonomous underwater vehicle, or AUV — might be to use so-called acoustic tethers. Since 1981, an ROV called Epaulard has logged well over 200 dives, 56 of them to depths of three miles or more. It communicates with the surface entirely by sound impulses. Made in France (epaulard means "killer whale" — an apt description of the vehicle's shape), the craft weighs 2.9 tons and is powered by a 48-volt lead-acid battery, enough for an eight-hour mission. Unhampered by clumsy cables, Epaulard has investigated an underwater volcano off the coast of Italy, prospected for manganese fields in the Mediterranean, surveyed and photographed hundreds of square miles of Pacific Ocean bottom, located a World War II Douglas Dauntless bomber at a depth of 4,200 feet off the coast of San Diego, and even snapped photos of a soft-drink bottle at 9,000 feet.

The next — and most exciting — step: a truly autonomous vehicle that will be able to navigate and survive the ocean depths on its own. The University of New Hampshire's Marine Systems Engineering Lab is approaching that goal with its Experimental Autonomous Vehicle-East, an underwater robot designed to inspect pipelines and the interiors of offshore structures.

EAVE-East, as the robot is known, is equipped with five microprocessors programmed with a description of its assignment and a kind of expert system consisting of decision-making rules and facts that could affect the mission. In response to information from its sensors, the robot can adjust power to its thrusters and change course if, for example, it is in danger of grounding.

Complex microcomputer programs for control, guidance and independent performance of practical tasks are still under development, and the challenges are formidable. "The biggest stumbling block," says D. Richard Blidberg, who has been overseeing the evolution of EAVE-East, "is programming judgment into the system. An AUV may easily handle virtually all expected and unexpected events as well as any combination of internal systems failures. But true human judgment and plain common sense are a long way off."

A DOCTOR'S AID

One arena in which human judgment and common sense are irreplaceable is the hospital operating room. Yet even here, robotic assistance has proved to be an unexpected boon: Robotic surgical tools are increasingly being used in place of surgeons' hands in extremely delicate operations.

The most notable medical robot to date is Ole, a mechanical neurosurgeon that assists California doctors. Ole performs jobs that give even the most skilled of human surgeons pause. When it comes to the brain, the less tissue a surgeon has to violate, the better. Every time a surgical instrument enters the brain, passing through blood vessels and other critical areas, there is an increased danger that the patient will suffer impaired vision or speech, reduced motor control or even paralysis. Before operating, a surgeon wants to know — with as much accuracy as possible — where inside the brain a suspected tumor or blood clot is. In the past, if the problem was very small and deep within the brain, doctors would forgo surgery altogether, believing the risk to be greater than the benefit. But the patient might die as a result.

An impressive piece of medical technology called a CAT scanner (for Computerized Axial Tomography) has alleviated the problem somewhat. CAT scanners are able to take three-dimensional X-rays of the brain and project them onto video monitors to reveal any abnormalities. But directing a surgical instrument into the area is a much less exact process. Since the late 1970s, the standard procedure has been to fit a patient under the CAT scanner with a head ring, or helmet-like frame, called a stereotaxis device. The calibrated frame superimposes a series of coordinates over the X-ray image, locating the problem area and indicating the path the instrument should take.

But it still remains for the surgeon to insert the hand-directed instruments at just the right point and angle, plunge to just the right depth and hope to score a bulls-eye. Doctors refer to this method of operation as freehand — not at all a comforting term for the person under the knife.

Dr. Yik San Kwoh, director of CAT scan research at California's Long Beach Memorial Medical Center, observed his neurosurgeon colleagues at their demanding work and began to think about robots. "I didn't know anything about robots," he said, "but I had seen pictures — on TV, I suppose — of robots working on car assembly lines. I thought, why not use them in brain surgery?"

So Kwoh, an expatriate from Shanghai who is neither robot scientist nor physician but electrical engineer, set out to develop a robot neurosurgeon. His first stop was the Yellow Pages, under "Robots." There was one company listed, and it provided robots for entertainment at birthday parties and openings of fast-food stands. For Kwoh, it was not an auspicious beginning. Eventually, though, his search led him to Unimation, Inc. in Danbury, Connecticut, one of the country's leading manufacturers of industrial robots *(Chapter 2)*. Unimation lent Kwoh one of its assembly-line robot arms. Soon he was trying to persuade the hospital to buy one and let him convert it for surgery.

An experimental robot was not among the hospital's priorities. But in 1981, an 81-year-old immigrant Dane named Sven Olsen donated $65,000 for the purchase of a robot arm; further gifts over the next several years would bring the total to $300,000. During a celebration of the hospital's new purchase, the 29-pound aluminum arm turned to Olsen and shook his hand, and Kwoh announced that the robot would be named Ole, in honor of its benefactor.

By January 1985, Kwoh had devised a guide tip for Ole's arm that would enable surgeons to target their instruments precisely, and a computer program that allowed the robot to work in conjunction with a CAT scanner. It remained only to test Ole's accuracy. Kwoh did so by pushing a pellet from a BB gun deep inside a watermelon and placing the melon under the CAT scanner. When the image of the pellet appeared on the video screen, he maneuvered a track-ball controller to guide the screen cursor to the spot. At Kwoh's typed-in command, the computer calculated the location of the pellet and relayed instructions to the motors for Ole's six joints. The motors whirred softly as the robot arm positioned itself above the watermelon. Then, when the computer indicated that the probe guide at the tip of the arm was in place, Kwoh gently pushed the needle down through the arm's guide. It hit the BB unerringly.

Ole's first human patient was a 52-year-old man with a brain lesion suspected of being a tumor. Surgeon Ronald Young located the area on the CAT scan, Kwoh typed a command into the computer, and Ole swiveled into position. Then, as Kwoh had done with the watermelon, Young pushed a biopsy needle through Ole's guide and directly into the tumor.

Ole has done numerous biopsies since. Because the robot arm is accurate to within 1/2,000 of an inch, one 77-year-old woman needed no more than a half-inch incision — conventional biopsy surgery can demand cuts inches long — and it was possible to use a local anesthetic. Two stitches closed her scalp, and with a small bandage over the wound, the woman went home the next day.

As successful as Ole has been, Kwoh envisions the robot doing more than brain surgery; it might be used to drain abscesses, for example, or to help orthopedic surgeons repair blood vessels, ruptured discs, torn ligaments and damaged carti-

Driverless Trucks
for Moving Materials

In warehouses and other facilities where heavy loads are constantly moved about from one place to another, more and more of the work is being shouldered by mobile computerized devices known as automated guided vehicles, or AGVs. Unlike robotic relatives that are bolted beside assembly lines *(pages 47-59)*, these driverless trucks travel from one assignment to the next by following designated paths on the floor *(pages 108-109)*.

As shown here, AGVs come in several versions, but all have certain characteristics in common. For example, they are designed to move slowly and are equipped with sensors and brakes to avoid collisions with humans or other objects. In many systems, an onboard computer enables each vehicle to plan its own route between specified stops, while a central computer is responsible for keeping track of the whereabouts of each AGV in the plant's fleet.

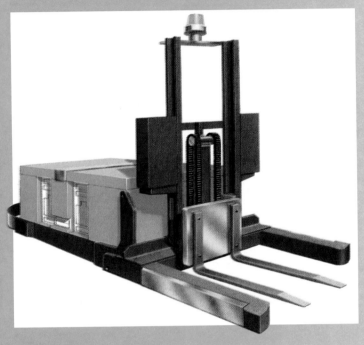

Each of the AGVs depicted here is a specialist. The guided fork-lift truck at left can stack or retrieve pallets or containers at heights up to 16 feet above a warehouse floor. The vehicle below, at left, is a mobile conveyor platform that can drive up to a fixed conveyor to pick up or drop off a standard-size package; the movement of the load onto or off the AGV's roller-top deck is controlled by its onboard computer. More useful for big payloads and trips of 200 feet or longer is a self-steering tow tractor *(below)*, capable of pulling a 20,000-pound train of several trailers. For safety, obstacle-sensing bumpers and side detectors watch for obstructions, and a constant bell and flashing light warn humans of this AGV's approach. The tractor is also equipped with several stop buttons for emergencies.

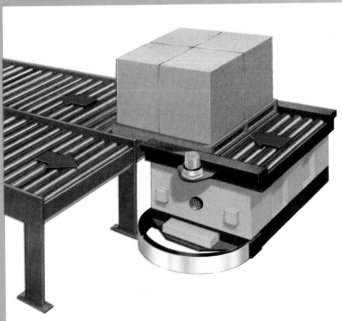

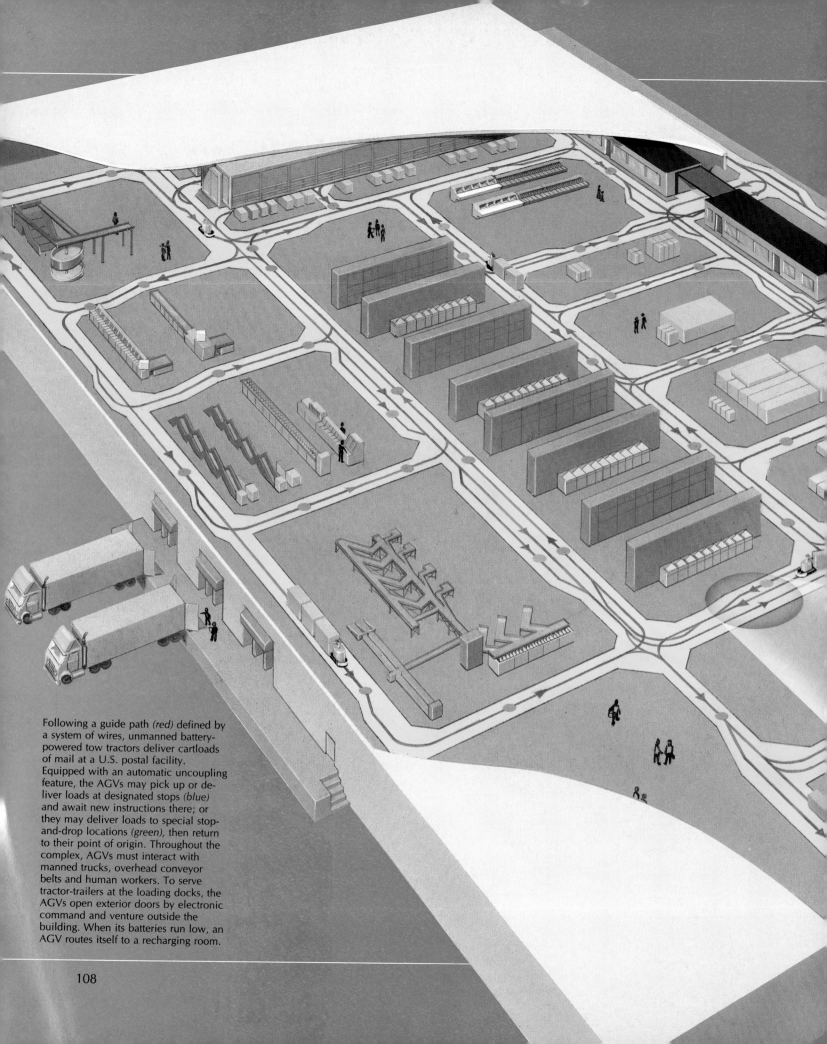

Following a guide path *(red)* defined by a system of wires, unmanned battery-powered tow tractors deliver cartloads of mail at a U.S. postal facility. Equipped with an automatic uncoupling feature, the AGVs may pick up or deliver loads at designated stops *(blue)* and await new instructions there; or they may deliver loads to special stop-and-drop locations *(green)*, then return to their point of origin. Throughout the complex, AGVs must interact with manned trucks, overhead conveyor belts and human workers. To serve tractor-trailers at the loading docks, the AGVs open exterior doors by electronic command and venture outside the building. When its batteries run low, an AGV routes itself to a recharging room.

Coordinating an Automated Fleet

Whether transporting bottles of Perrier in France, dairy products in Holland or, as shown here, mail in the U.S., AGVs systematize the flow of materials through a plant. Spread over six acres, the workroom floor of the typical postal facility shown here relies on a fleet of AGVs to help unload, process and ship out millions of pieces of mail daily. Managing AGV traffic in a maze of sorting machines, conveyor belts and human workers requires complex programming. Like a taxi dispatcher, a central computer monitors the location of each AGV in the facility and decides which ones should respond to calls for pickups. Once a vehicle has been paged, its own onboard computer determines the most efficient route to the destination. At the pickup stop, a human worker simply keys in the AGV's next destination or instructs the vehicle to deliver the load and return for another.

AGV routes, normally laid out as one-way paths, often consist of guide wires embedded in shallow slots in the floor. Each wire emits a particular frequency, designating a different complete loop through the system. The AGV's guidance sensor selects the appropriate guide wire for its destination. Avoiding collision with other AGVs — a process known as blocking — can be effected through the use of zones (below).

AGV RULES OF THE ROAD

Collisions between driverless vehicles are prevented by one of two techniques, called zone control and forward-sensing control. With forward-sensing control, used only on straightaways with no intersections, the AGV employs optical, sonic or other sensors to detect the presence of a vehicle ahead.

With zone control, AGV pathways are divided into zones (diagram) that have electronic or magnetic markers embedded at the boundaries. AGVs are admitted to a zone one at a time on a first-come, first-served basis. Each AGV continuously reports the zone it is in and monitors the locations of other AGVs. In some systems, a central computer determines priorities, but in many — including the system shown at left — AGVs exercise self-control. Here, for example, if AGVs from Zones B and C need to enter Zone A, and the one from Zone C arrives first, the computer aboard the AGV in Zone B would order it to wait until Zone A is empty. If two vehicles arrive at a zone boundary simultaneously, certain rules of priority take effect. For instance, the AGVs' computers may be programmed to allow vehicles entering from a main trunk line to take priority over AGVs entering from a spur.

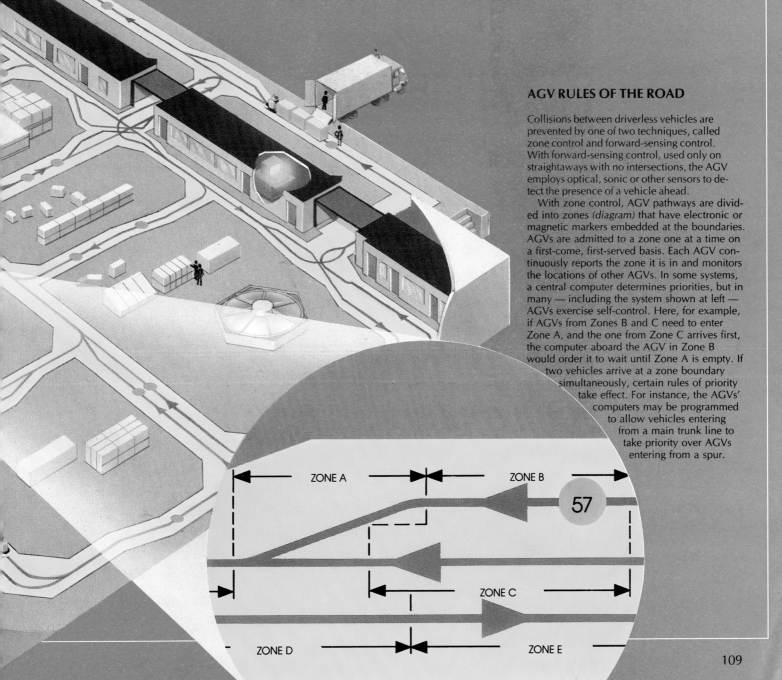

ZONE A   ZONE B

57

ZONE C

ZONE D   ZONE E

lage. And he is not concerned about competitors invading a field he originated. In the spirit of "Ole" Olsen, who died in 1984, Kwoh is eager to share his discoveries. "Ole always told me that nothing made him more happy than to help others," he said. "I want to keep everything that way."

BREAKING THE FIRST LAW

Ole the robot neurosurgeon is an ideal embodiment of writer Isaac Asimov's fictional First Law of Robotics: A robot may not injure a human being. But life is not fiction, and like most advanced technologies, robots have been used to do harm. Their military potential is immense.

On May 14, 1982, the British destroyer *Sheffield* was on radar picket duty in choppy waters about 70 miles north of the Falkland Islands. H.M.S. *Sheffield* was a modern, computerized, 4,000-ton fighting ship, protected by some of the most advanced defensive systems available. But twenty miles away, a distance too great to allow him to see his prey, an Argentine pilot pointed his French-built Super-Etendard fighter-bomber in the general direction of the ship. He pushed the launch button and turned for home. Less than three minutes later, an Exocet missile struck the *Sheffield* amidships. "We had time only to say 'Take cover,' " the destroyer's commander, Captain James Salt, remembered afterward. "Three or four seconds later the missile hit." It penetrated to the center of the ship before its 360-pound warhead exploded, igniting the rest of its volatile propellant. "On the upper deck you could feel the heat through your feet with shoes on," Salt said. "The superstructure was steaming, and the paint on the sides was coming off. The hull was glowing red and hot. We had no hope of retaining the fighting capability of the ship." After hours of futile battle against the flames, Captain Salt gave orders to abandon ship, thereby saving all but 20 of his crew.

The Exocet missile that caused such devastation was a robot. It was 15½ feet long, weighed 1,440 pounds and could be launched from an airplane flying as high as 33,000 feet or as low as 300 feet. The Argentine pilot needed only to program the missile's computer with the *Sheffield's* range and bearing. Once he released the missile, it flew at more than 500 miles an hour, six to eight feet above the water — too low to be detected by defensive radar. The computer kept the missile on course, responding to impulses from various sensors and gyroscopes. At about seven miles from the British destroyer, an active homing radar took over. Rising slightly to scan the horizon, the missile made sure it was locked in on target. (If the *Sheffield* had shifted its position in the meantime, the Exocet's computer would have adjusted its course.) Then it descended to near wave level again until, like a pilotless Kamikaze raider, it struck. The men on the *Sheffield* had no idea what hit their ship.

In the Falklands war, six Exocets were fired by Argentina; four damaged or destroyed their targets. In all, of the 10 ships and 114 aircraft lost by both sides, more than half were destroyed by flying robots. These so-called smart weapons have more than a 90-percent chance of striking their targets — a rate previously unheard of — and in comparison with their targets they are laughably cheap to produce: The Exocet missile cost $200,000, the *Sheffield* $50 million.

According to some military thinkers, fighting robots like the Exocet have shifted the advantage in war to the defense. In order to win territory, offensive armies must expose their valuable hardware to increasingly intelligent and lethal defensive weapons. For example, during the Sinai campaign of the 1973 Arab-Israeli war, the 190th Israeli Brigade attacked the Egyptian Second Army with a massive force of tanks protected by an umbrella of air support. Within a few hours, 130 Israeli tanks were destroyed by Russian-made computer-directed Sagger missiles, which weighed only 25 pounds each. And not only the Israeli tank brigade was annihilated — so too was the traditional notion that, in war, sheer mass of firepower spells success. Increasingly the balance seems to be tipping in favor of "intelligent" single-shot weapons, which do their work with deadly efficiency. One of the most formidable of such weapons is the U.S.'s Tomahawk cruise missile, a fighting robot designed to seek out a target up to 1,500 miles away and fly to it at close to the speed of sound. Once in the vicinity of its target, it may make last-minute evasive twists and turns before striking. The Tomahawk can carry either a nuclear or a conventional warhead, making it capable of controlled strikes against strategic targets without destroying whole cities and huge civilian populations. In one test of its accuracy, it was launched from a submarine off the coast of California, flew 300 miles inland and ripped in two a cloth banner stretched between two poles. One expert has called it "a killer riding a computer."

In time, these devastatingly effective computerized killers will be tended by a host of other robots, each designed to do a job that formerly required people. The Futures/Long-Range Planning Group of the U.S. Army War College's Strategic Studies Institute imagines a battlefield scene in which robots identify and track enemy aircraft and other targets, detect mine fields, implant "smart" mines and build obstacles to protect their human comrades. Robots also may load weapons, transfer ammunition from storage areas to the front and take care of routine administrative chores, such as issuing uniforms and equipment.

This kind of fighting and support capability, combined with developments in such other robot weapons systems as tanks, walkers and crawlers, has prompted one engineer to write that "in the distant future, there may be no human soldiers at all on the battlefield. Instead, train carloads of military robots may be shipped to the battleground to engage in modern mechanized jousting, to ascertain, by proxy, which of their sponsors should be declared the ultimate victor."

SOLAR SYSTEM SCOUTS

More peaceful automated surrogates have been operating on the frontier of space for nearly two decades. Since 1976, human beings have been sending robotic probes to outer worlds as scouts and pioneers. In that year, two Viking landers descended on Mars. Besides the ability to transmit photographs — which have since made the red face of Mars as familiar to planetologists as the face of earth — each lander possessed a robot arm that could stretch eight feet from its base, extend a shovel and scoop up samples of Martian soil, which it then dropped into a hopper on top of the spacecraft. The hopper led inside to three automated analysis laboratories and growth chambers.

The Vikings searched — in vain, as it turned out — for life on the Red Planet, testing first to see if anything in the soil took up carbon from the air, as plants do on earth to synthesize nutrients. In another experiment, the spacecraft fed a few drops of nutrients to the soil samples to see if anything would consume the food, and in yet a third, they added water vapor to see if the soil would produce oxygen.

The landers also measured the depth and composition of the crust on Mars and the planet's weather patterns. Altogether, the Viking landers worked without fail for more than two years, until they finally lost radio contact with earth.

The Galileo mission to Jupiter is designed to achieve even more. Launched on October 18, 1989, it has the highest level of autonomy of any space mission ever flown. The spacecraft consists of two entities: an orbiter to undertake a 20-month exploration around Jupiter and many of its moons, and a bullet-like probe that will plummet from the orbiter into Jupiter's clouds. The probe will deploy a parachute and conduct a slow, controlled exploration, until the immense planet's crushing atmospheric pressure — 10 times that on earth — silences it forever.

At a cost of close to a billion dollars, Galileo is nothing less than a two-and-a-half-ton, automated, self-contained scientific laboratory. It carries with it 11 instruments designed to study the composition of both Jupiter's atmosphere and its interior. It will also study the chemistry and mineralogy of the planet's moons, flying so close to some of them that features as small as 90 feet across will be clearly visible.

The programming required to achieve these goals is equally impressive. For example, since the probe is unable to maneuver by itself, it must be released from the orbiter at exactly the right moment to allow it to enter Jupiter's atmosphere within an imaginary window less than 500 miles across; moreover, its angle of entry cannot vary by more than three degrees. If the probe is minutely off course, it will bounce off the atmosphere like a stone skipping off the surface of a pond and head out into space. The release point is more than 93 million miles away from the planet — calling for navigational accuracy comparable to striking a target one foot in diameter with a bullet fired from 25 miles away.

After Galileo, which is scheduled to arrive at Jupiter in late 1995, the possibilities are infinite. NASA has outlined plans to exploit earth's moon by means of robots. Unmanned solar-powered rovers will explore the moon's surface, then prepare for a manned lunar base supplied by an automated factory. Eventually, according to NASA's scenario, another automated factory will use the resources of the moon to produce consumer goods — including household robots — for earth and for pioneer settlers on Mars. Perhaps the factory's most significant product will be intelligent robot spacecraft equipped with vision and tactile sensory systems. Through these robot explorers, earthbound voyagers will gain vicarious experience of space travel.

Finally, experts predict, robots will be able to autonomously reproduce. When that day comes, the stars themselves will no longer be too remote for exploration. Automated craft will replicate at each solar system, plunging deeper and deeper into the vastness of the universe. Whatever the end of such explorations, one thing seems clear: It will be robots that lead the way.

In a more immediate and mundane future — one that has already started to impinge upon daily life — robots will be changing the way people live on earth. Some experts foresee widespread applications of robotics to workaday tasks — so many that tomorrow's society might, in the words of one futurist, be viewed as "an Athens without slaves." For example, robot construction workers, adaptations of existing factory robots, could eventually free human workers from at least some of the labor and the hazards involved in building homes, bridges and roads. Similar software and sensory systems might also be applied to the needs of householders. In both Japan and the United States, researchers are investigating the potential of robotic nurses. The Japanese have developed a robot orderly that can lift a bedridden patient out of bed and carry him or her to the X-ray or operating room. In the U.S., robots capable of feeding, shaving and tending to the general needs of quadriplegics are serving the Veterans Administration in Palo Alto, California. Paralyzed patients direct their robotic health-care attendants with simple spoken commands.

Similar technology may one day be employed for quite another purpose Down Under: to shear sheep. Since the late 1970s, several Australian firms have been attempting to create an automated shearer capable of quickly gathering wool without hurting the sheep. The most advanced of these projects is a two-armed robot developed by Merino Wool Harvesting of Adelaide. The device, which has sensors that provide feedback to steer its cutters over the contours of the animal's body, is designed to give the sheep a clean shave in just 90 seconds. At that rate, the robot could shear 300 sheep a day — twice the quota of a human cutter.

Agricultural engineers in the United States are working on robotic orange pickers that can sense when the fruit is ripe and robotic weeders that are able to distinguish food crops from weeds. Thinking bigger, roboticists such as James Albus of the National Institute for Standards and Technology have speculated that machines may also alleviate some of our more pressing long-term problems, including the depletion of energy sources.

ISLANDS OF ENERGY

Albus envisions such grand projects as floating islands, resembling lily pads up to six miles in diameter, which would be used for farming algae and processing it into fuel-grade alcohol for energy use. These plastic-bottomed ponds, filled with highly fertilized sea water and kept afloat by a web of tubing, would drift on the surface of the equatorial seas. They would use the movement of the waves to keep the algae in constant circulation from the outer portions of the pad to the processing region in the center. The entire operation would depend on a network of microcomputers to adjust small sails and rudders that would keep the structure seaworthy. The microcomputer control center could even allow for the huge floating still to fold up and sink during rough weather and to rise and spread open again when the storm has passed.

Albus expects research in genetic engineering to increase the efficiency of the algae-to-alcohol process to the point that 2,500 lily pads would be able to produce the equivalent of all the fuel oil consumed in the United States. According to Albus, "The alcohol fuel available from robot lily pads might eventually be sufficient to provide the entire energy needs of the world for the foreseeable future" — and without any pollution as a by-product.

The robot as gigantic lily pad is an enchanting concept, and possibly even practical. It may never exist other than as a dream in the minds of those who seek knowledge, beauty, excitement and delight — but then, neither would QN have existed in its modern form if someone had not been tantalized by an idea.

Perhaps the most fascinating robot of all, QN now hangs in the Smithsonian Institution's National Air and Space Museum. Also called the Time Traveler, this robot is a replica of a flying dinosaur, *Quetzalcoatlus northropi,* which has been extinct for 65 million years. A pterosaur with a 36-foot wingspan, it was the largest animal ever to take to the skies. But according to the laws of aerodynamics and biology, it could not have flown at all. It was simply too big.

Among those intrigued by the problem of a flying impossibility was Paul MacCready, an aeronautical engineer who heads a research group called Aero-Vironment, near Pasadena, California. In 1977, MacCready and his team designed and built the *Gossamer Condor,* the first human-powered airplane to fly a controlled course. Two years later, his *Gossamer Albatross,* also human powered, crossed the English Channel, and in 1980 the *Gossamer Penguin* became the first solar-powered plane to make a climbing flight. MacCready was used to unusual challenges; QN would be one of the most demanding. With $500,000 of funding from the Smithsonian and the Johnson Wax Company, he set out to discover how the original QN flew — by building a working robot model.

It was not easy. Fossils showed that the pterosaur had no tail, which meant that it probably used its large head and long beak as a vertical rudder — a task as difficult, according to one of MacCready's designers, as "trying to fly an arrow with the tail feathers in front" — and it controlled pitch by sweeping its wings forward and back, constantly changing its center of gravity. The robot would have to monitor the wind's effects on its head and wings and make constant adjustments — all while flapping away hard enough to stay aloft in the first place.

Little by little, MacCready's team progressed from a gliding model with an eight-foot wingspan to an 18-foot version. At the same time, they learned how to do without a tail. The next challenge was getting the wings to flap realistically. Eventually, the engineers combined flapping with movement of the robot's clawed fingers, which can extend slightly to create drag and help QN to turn. They also allowed the wings to twist, providing lift that brings one wing up while the other drops, further helping the great creature to navigate. Finally, they fitted QN with wind sensors affixed to its neck, gyroscopes to monitor turning speed, an onboard computer to process the information and six pounds of nickel-cadmium batteries to provide the 35-pound robot enough power for five minutes of flight. On January 7, 1986, over a dry lake near Edwards Air Force Base in California, QN flapped its wings, swiveled its head, jettisoned the tail boom that gave it stability on take-offs and glided into the twilight sky. It subsequently flew perfectly 21 times in a row.

Still to come, however, was its public unveiling. On May 17, in view of thousands attending an air show at Andrews Air Force Base outside Washington, D.C., QN took to the air once again. This time the unthinkable happened: Just seconds after flying free from a towline that had pulled it to a height of 400 feet, the great bird caught a gust of wind, dropped its head toward the earth and crashed, breaking its neck in the process. MacCready's engineers explained afterward that the dive had been caused by QN's tail boom dropping off too soon, and even though the robot made a brief recovery after going on auto-pilot, the force of the wind against the mechanical pterosaur during its dive had cracked the supports between its head and neck.

After 65 million years, the dinosaur known as *Quetzalcoatlus northropi* had flown again and was again extinct. For an instant, a robot had taken 20th-century citizens back in time, the newest of human technologies illuminating the most distant past, demonstrating once again both the enormous reach of humanity's creations and their limitations.

Keeping Danger at a Distance

In the world beyond the assembly line, robots are often called upon to perform tasks too hazardous for human beings. Because the rote motions of a manufacturing robot would be inadequate for unstructured settings, however, engineers in many fields have built human supervision into the control systems of the robots. Such remotely operated devices, called teleoperators, draw on human intelligence to apply their robotic muscle.

Designed for endurance, strength and reliability, a teleoperator can be equipped with any number of items from the catalogue of robotic hardware — including wheels or treads to trundle it to the site, grippers or saws to execute the work, and video cameras or force sensors to report its accomplishments to the human controller. At a safe remove, the teleoperator's human partner contributes general skills that so far have resisted computerization — interpreting events, learning from mistakes, deciding among alternatives.

When computers are used in teleoperation, they tend to play a supporting role. Typically, a computer might convert the human operator's movements into digital commands, transmit control signals to the teleoperator, monitor the machine's performance and translate sensory data collected by the teleoperator into feedback for its controller. These custodial services provide valuable refinements in a teleoperator's feedback loop *(pages 12-13)*, where the accuracy of the operator's commands hinges in part on the clarity of the feedback the robot is able to furnish.

The following pages present a gallery of teleoperators with progressively sophisticated control devices and feedback abilities. As the sequence shows, teleoperation systems are evolving toward controls that increase the immediacy of human responses to whatever circumstances the machines encounter. Efforts to hone the robots' sensory faculties, meanwhile, have focused on developing ever-more vivid visual displays, for sight conveys more pieces of information simultaneously than any other sense.

Salvage Operations by an Undersea Servant

Remote operation is ideal for work beneath the sea, where darkness, high pressure and low temperatures permit only brief, perilous forays by humans. A teleoperator powered and controlled from the water's surface can stay submerged for weeks at a time, performing tasks that range from exploration to retrieval *(pictured below)* to assembly.

Manipulating such an underwater marionette is an exacting science. In addition to navigational guidance, the remotely

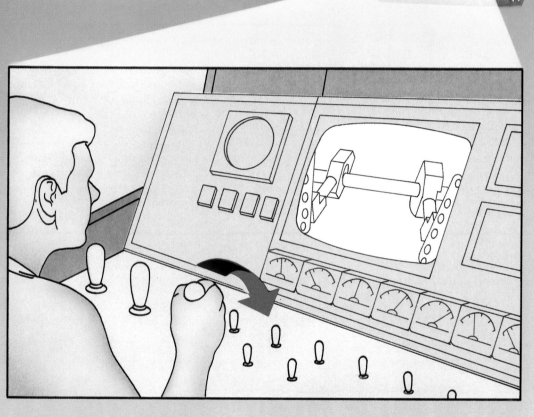

Directing the movements of a submersible ROV *(right)* from the shipboard control panel *(above)*, an operator maneuvers a joystick to clamp the robot's left-arm rotary saw onto a damaged length of undersea telephone cable. The robot's right-arm gripper steadies the cable as it is severed, while the operator's video screen and a bank of gauges monitor the work in progress. Later, the robot will affix lines to the cable so that the cut ends may be hauled to the surface for repair by the ship's crew; the operator will then instruct the robot to rebury the spliced cable.

operated vehicle (ROV) requires detailed commands to maneuver its arms. The simplest systems assign a control switch to each of the ROV's "degrees of freedom," or directional movements. A jointed wrist, for example, would be operated by three switches governing three types of motion: vertical, horizontal and rotational (also known as pitch, yaw and roll). In more advanced systems, joints that require greater speed and coordination may be manipulated by joysticks: With a single thrust of the joystick, the human controller can move an entire joint, dictating the speed and extent of the movement as well as its direction.

As if switches and joysticks were not a crude enough means of control, the feedback the operator depends upon is often frustratingly sparse. Along with readings from analog gauges, the feedback may consist of nothing more than video images of segregated portions of the work site, distorted by the vehicle's wide-angle camera lenses and illuminated solely by the vehicle's own lights. As optical fibers enjoy wider use, the wealth of data passing between control ship and ROV will allow for a tighter mesh between the images and other feedback the machine reports, and the commands its human operator issues in return.

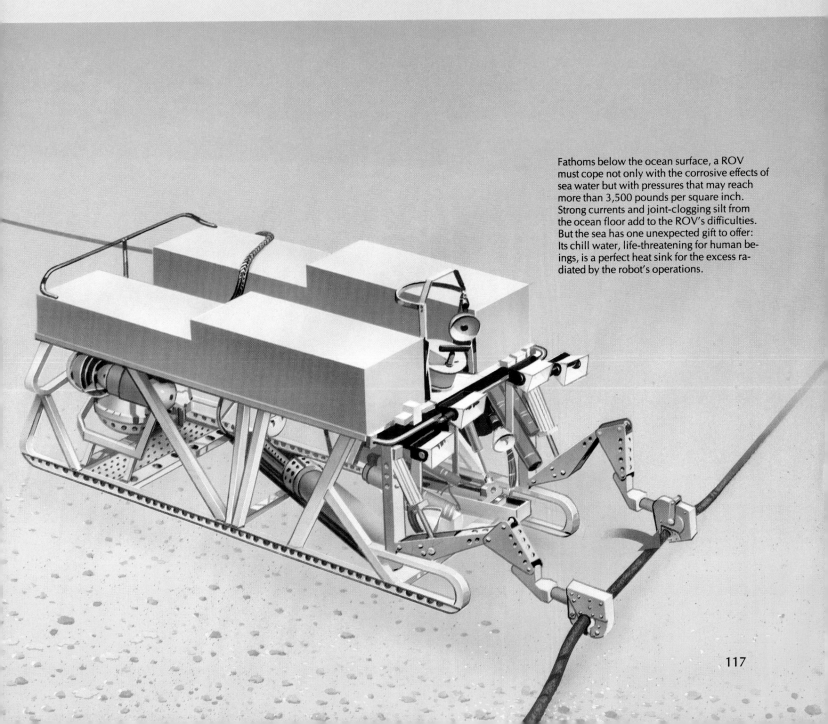

Fathoms below the ocean surface, a ROV must cope not only with the corrosive effects of sea water but with pressures that may reach more than 3,500 pounds per square inch. Strong currents and joint-clogging silt from the ocean floor add to the ROV's difficulties. But the sea has one unexpected gift to offer: Its chill water, life-threatening for human beings, is a perfect heat sink for the excess radiated by the robot's operations.

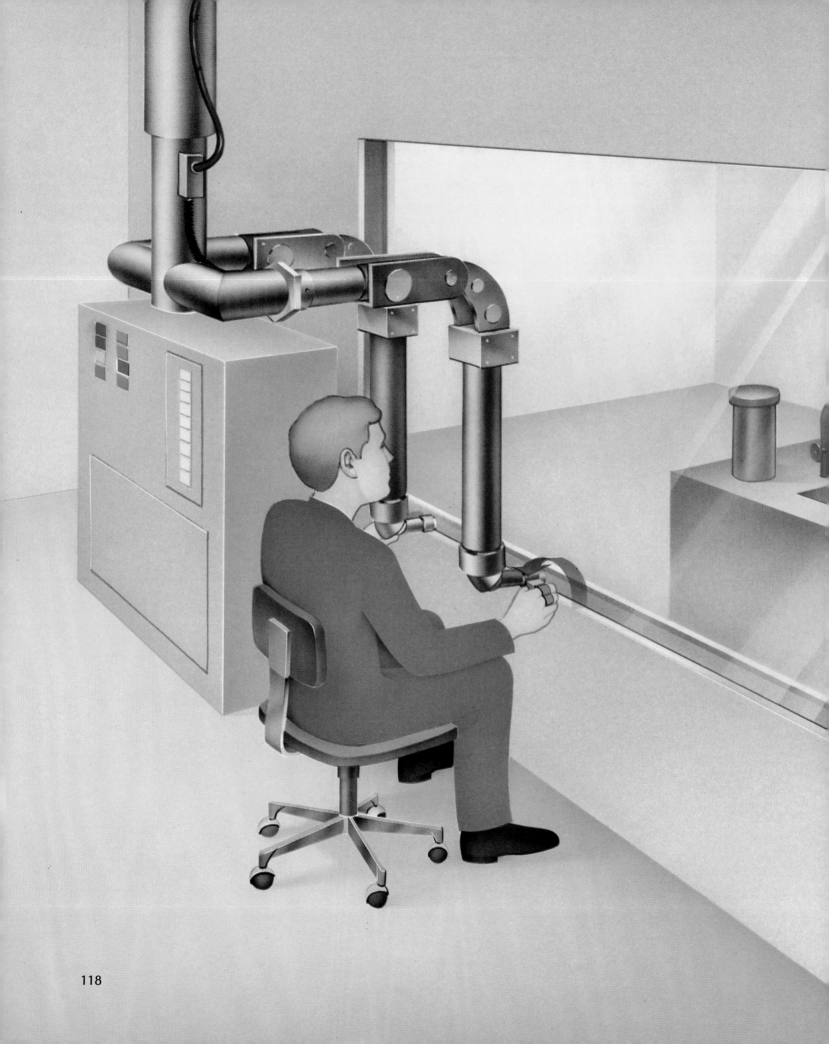

Master-Slave Labor in a "Hot" Lab

Although switches and joysticks characterize many teleoperated systems, the meticulous maneuvering of robotic joints can fatigue a human operator in quick order. Where feasible, so-called master-slave controls, in which a "slave" manipulator precisely mimics the operator's control mechanism, are often used instead. As the operator works the controls, he or she watches the slave's resulting movements through a protective window or on a video screen. With force feedback — the replication in the master controls of physical resistance experienced by the slave — the operator can gain a realistic feel for the stresses the machine encounters.

Master-slave controls appear most frequently in radiation-research labs, where the devices first came into use; often, these devices are not computerized. In some nuclear applications, radiation rates may exceed 10,000 rems per hour, a dosage instantaneously fatal for humans. Such heavy exposure can threaten the health of the teleoperator machine itself. In much the same way that solar radiation causes sunburn, extremely high rates of nuclear radiation can quickly age normal materials. These exposures wreak special havoc on organic substances such as hydraulic and lubricating oils; consequently, slave manipulators rarely function hydraulically and often they feature dry lubricants.

Introducing a computer to the master-slave equation can yield several bonuses. In cases where the distance between man and machine cannot be bridged mechanically, a computerized master device can animate a slave manipulator miles away — deep in the sea, for example, or deep in space. Some computer-aided systems protect the teleoperator from an overzealous master with programmed rules restricting the extent of the robot's motions. An operator's gestures may even be amplified for greater strength, or scaled down for delicate work.

Relying on the spatial correspondence between his master controls and the slave teleoperator, a laboratory technician pours radioactive waste from a test tube into a beaker. The lead-glass window enables the technician to see into the "hot" room where the experiment unfolds; unlike the simplified window shown here, however, laboratory glass normally measures more than three feet thick — less than ideal for observation, but substantial enough to shield the operator from radioactive hazard.

An Exoskeleton for Ultimate Control

In many systems run from afar, the operator's ability to mentally project himself into the work site will be the crux of accurate control. Like the writing in a good novel, the control and feedback in such systems should convey an immediate sense of "being there," becoming almost unnoticed as the operator concentrates on taking part in the scene.

Researchers working to enhance the operator's telepresence, as this sensation is known, have developed exoskeletal controls, which adhere snugly to the operator's body. As the

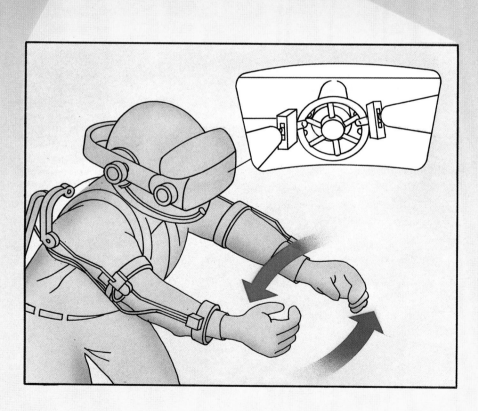

Controls on an operator's head, arms, hands and torso translate body motions (left) into movement commands for a teleoperator, which responds more than a mile away by shutting off a noxious-gas leak from a ruptured pipe joint (right). The teleoperator's twin video cameras transmit a stereo image to a display mounted inside the helmet (left, inset), allowing the operator to react to events in the gas-filled chamber. As the cost of teleoperators decreases and the machines become easier to control, they may play an increasingly helpful role in remedying such mishaps.

exoskeleton records the body's movements, a support computer converts them into commands that are flashed to a humanoid teleoperator. Because the exoskeleton encodes the separate motions of head, arms, hands and torso, the operator requires very little training in teleoperator control. The intuitive ease of this method may one day be magnified by such feedback as three-dimensional color displays and tactile sensation.

Proponents of anthropomorphic teleoperators — remote-controlled robots designed to resemble humans — stress their versatility as much as their ease of use. Because these teleoperators are modeled after the human body, they are theoretically capable of performing any work humans can do. Since teleoperators with nonhuman shapes and abilities remain essential for many tasks, however, the control systems outlined in the preceding pages will likely be used for years to come. Telepresence will complement, not co-opt, existing controls.

Glossary

Actuator: a device that converts electrical, hydraulic or pneumatic energy into robotic motion.

Analog: the representation of a continuously changing physical variable (pressure, for example) by another physical variable (such as electrical current).

Anthropomorphic: having the physical form, or other qualities, of a human being.

Arm: an industrial robotic manipulator made up of a series of rigid links and powered joints that terminates in a wrist socket, to which specialized tools, or end-effectors, can be attached.

Artificial intelligence: computer programs that emulate such human abilities as learning, perception and reasoning.

Automated guided vehicle (AGV): a driverless truck that picks up and delivers materials, usually within a building.

Automation: the operation and control of machines or processes without direct human supervision.

Autonomous: capable of independent action.

Bang-bang robot: a simple industrial robot arm, usually controlled by fixed mechanical stops rather than by feedback, that transfers materials from one place to another; also called a pick-and-place robot.

Batch manufacturing: production of parts or materials in discrete runs or lots, interspersed with the production of other parts; not continuous, as in mass production.

Binary: having two components or possible states, usually represented by zeros and ones; in a digital computer, information is processed and stored in binary form.

Binary image: an image composed of only two brightness levels (black and white).

Bionics: the application of biological principles to the solution of engineering problems.

Cartesian coordinate: any of three coordinates — usually designated X, Y and Z — that locate a point in space and measure its distance from any of three intersecting coordinate planes. In robotics, the coordinates are used to identify points for the positioning of an end-effector.

Cartesian-coordinate robot: a robot arm with three linear joints; its position is expressed in Cartesian coordinates. Also called a rectangular-coordinate robot.

Center of gravity: the point at which the entire weight of a robot or any other body could be considered concentrated, so that if supported at this point the body would remain in equilibrium no matter what its position.

Circuit board: a plastic board on which electronic components are mounted. Robots are increasingly used for installing circuit-board components.

Closed-loop control: a method of controlling robots in which the position of a moving member or the status of some other output is fed back to the control unit. *See also* Feedback loop; Servocontrol.

Coaxial cable: a transmission medium composed of an insulated conductive wire inside a tubular conductor.

Computer-aided design (CAD): the use of computers to shorten the design cycle by allowing manufacturers to shape new products on a computer screen without having to build physical prototypes.

Computer-integrated manufacturing (CIM): the use of computers to control the flow of information and materials through a factory; the term CIM implies a more comprehensive role for computers than does computer-aided manufacturing (CAM).

Continuous path: a type of robotic control in which a stored program specifies so many points along a desired path that the path is virtually continuous; contrasted with point-to-point.

Control: the process of matching a robot's actual performance with its desired performance. Also refers to a device a person may use to command a robot or a teleoperator.

Control unit: a computer that controls an industrial robot; also called a controller.

Cutaneous sensing: a type of tactile sensing that occurs at or near the point of contact with the sensed object.

Cybernetics: the study of control and communications systems in both machines and animals, founded on the theory that intelligent beings adapt to their environments and accomplish goals by reacting to feedback from their surroundings.

Cylindrical-coordinate robot: a robot arm with two linear joints and one rotary joint; its position is expressed in cylindrical coordinates, which define the location of any point in terms of an angular dimension, a radial dimension and a height from a reference plane. (These three dimensions specify a point on a cylinder.)

Degree of freedom: one way in which a robot can move. For example, most robotic wrists have three degrees of freedom: vertical, horizontal and rotational.

Digital: pertaining to the representation, manipulation or transmission of information by discrete signals.

Digital computer: a machine that operates on data expressed in discrete form rather than the continuous representation used in an analog computer.

Digitize: to represent data in digital, or discrete, form; or to convert analog, or continuous, information into discrete signals for use by a digital computer.

Downtime: a period when a system, such as a robot or a workstation, is unavailable for production.

Dynamic balance: in computerized walking machines, the active avoidance of falling, achieved by continuous adjustments to posture, velocity and elevation above the ground. Also known as dynamic stability.

Edge detection: the process of recording places in a digitized image where brightness values change abruptly; such locations usually correspond to an edge.

End-effector: a gripper or other tool attached to the wrist of a manipulator arm.

Exoskeleton: an articulated mechanism whose joints correspond to those of a human arm or other parts of the human skeleton; when placed around a human operator, it moves as the operator does. Exoskeletal controls may be used in teleoperation.

External sensor: a sensor that detects information about the robot's external environment rather than about the robot itself.

Feedback: information sent to a robot's control unit from internal or external sensors to convey either the robot's actual response to a command or the external effect of a robot's action.

Feedback loop: a control system incorporating controller commands to a mechanism, the mechanism's response, measurement of the response and transmission of the measurement back to the controller for comparison with the original command.

Fiber optics: the technology of encoding data as pulses of light beamed through ultrathin strands of glass or other transparent material.

Frequency: the number of times per second that a wave cycle (one peak and one trough) repeats.

Gait: a patterned sequence of leg and foot positions.

Gantry robot: a robot suspended from an overhead frame known as a gantry, along which it can propel itself.

Guided missile: a self-propelled, computer-controlled explosive device that can modify its course after launch, responding either to external direction or to direction originating from devices within the missile itself.

Hydraulic actuator: a device that uses pressurized fluid to produce linear or rotary motion.

Joint: the movable part of a robot that permits one link to rotate or slide relative to another.

Jointed-arm robot: a robot arm with three rotary joints; its position is expressed as three angular coordinates. Also called a revolute-coordinate robot.

Joystick: in teleoperated machines, a movable handle that is linked through a computer to the robot and used by a human operator to control the machine's movement.

Kinesthetic sensing: a type of tactile sensing that takes place away from the point of contact with an object, typically at the hand and arm joints, to detect characteristics such as weight.

Lead-through method: a means of programming a robot by physically guiding it through desired actions, often at reduced speed.

Light-emitting diode (LED): a semiconductor device that emits light when a current passes through it.

Link: a rigid beam between joints in a robotic arm.

Machine vision: computer perception of the external world by the interpretation of optical data.

Manipulator: a device, such as an industrial arm, composed of segments that rotate or slide relative to one another, for the purpose of grasping and moving objects; it may be controlled by a computer or a human being.

Mass production: continuous, large-scale manufacture of parts or material; contrasted with batch manufacturing.

Master-slave control: a teleoperation method in which the human operator's manipulation of a master control is mimicked by a similarly shaped slave device.

Memory: the principal area inside a computer into which data is recorded or from which it is retrieved; contrasted with storage, which is external to the computer.

Off-line programming: developing a written program to define a robot's desired motions. The program is then loaded into the robot's computer control unit for automatic execution.

Open-loop robot: a rudimentary robot that does not employ feedback; also called a fixed-stop or nonservo robot.

Optical processing: the direct manipulation of optical information without conversion to digital form.

Photoelectric sensor: a device that detects the presence or absence of light above a given threshold.

Pick-and-place robot: see Bang-bang robot.

Polar-coordinate robot: a robot arm with one linear and two rotary joints; its position is expressed in polar coordinates (one linear, two angular). See also SCARA.

Pneumatic actuator: a device that uses pressurized air to produce linear or rotary motion.

Point-to-point: a type of control in which a stored program specifies crucial points and the robot's control unit determines the path to be followed between points; contrasted with continuous path.

Potentiometer: a mechanically variable resistor that can control electrical voltages.

Program: a sequence of instructions for performing some operation or solving some problem by computer; in industrial robotics, for example, a program typically controls a robot's motion.

Programmable: capable of responding to instructions and thus able to perform a variety of tasks.

Prosthetic device: a device that substitutes for missing or impaired human limbs.

Quality control: inspection of manufactured products to determine whether they adhere to established standards.

Real-time: pertaining to computation that is synchronized with a physical process in the real world.

Rectangular-coordinate robot: see Cartesian-coordinate robot.

Remotely operated vehicle (ROV): a teleoperated vehicle with its own means of propulsion; underwater ROVs, for example, usually employ electric propellers.

Resolver: a type of internal sensor that measures joint positions by means of the relative placement of its movable and stationary electromagnetic components.

Robot: a programmable device capable of manipulation, locomotion or other work under computer control.

Robotics: the science of designing, building and using robots.

SCARA (selective compliance assembly robot arm): a robot with two rotary joints and one linear joint, in which both rotary joints are in the horizontal plane and the linear joint moves vertically; contrasted with a polar-coordinate robot.

Sensor: an information-pickup device that converts physical energy such as temperature or light into electrical signals, which may then be translated for use by the computer.

Servocontrol: a feedback control method in which the actual position of a robot's joints and links is continually compared with the desired position, and appropriate corrections made.

Servomechanism: an actuator or motor driven by servocontrol.

Servo valve: a device used in hydraulic and pneumatic actuators to control the flow of fluid through the actuator.

Signal: the information, often in electrical form, conveyed from one point (such as a computer) to another (such as an actuator).

Software: instructions, or programs, designed to be carried out by a computer or robot.

Speech recognition: the computer interpretation of human speech.

Static stability: in walking machines, locomotion in which the machine keeps a stable base of support on the ground at all times and maintains its center of gravity over those legs.

Stereo vision: a means of perceiving three-dimensional shapes based on the disparity between two images taken simultaneously from different positions.

Storage: a device that can receive, store and retrieve data, usually in binary form; storage devices used in robotics include cassette tapes, magnetic disks and so-called magnetic-bubble memory.

Structured light: light projected in a geometrical pattern to assist a machine-vision program.

Tactile sensing: the detection, through physical contact, of an object's characteristics.

Teach pendant: a hand-held control with which a robot may be programmed for point-to-point operation.

Teleoperation: the remote control of a robot-like device in which the primary controller is a human operator rather than a computer.

Walking machine: a robot or teleoperator that is capable of legged locomotion.

Work cell: an area consisting of one or more workstations and including storage and material-transfer facilities.

Work envelope: a three-dimensional space that includes only the points to which a robot arm can move its end-effector.

Workstation: an industrial unit consisting of a robot and the equipment with which it works.

Wrist: the joints between a robot arm and its end-effector that allow flexible maneuvering of the end-effector.

Bibliography

Books

Albus, James S., *Brains, Behavior, and Robotics*. Peterborough, N.H.: BYTE Books, 1981.

Asimov, Isaac:
Asimov on Science Fiction. New York: Doubleday, 1981.
I, Robot. New York: Fawcett, 1950.

Asimov, Isaac, and Karen A. Frenkel, *Robots: Machines in Man's Image*. New York: Harmony Books, 1985.

Ayes, Robert U., and Steven M. Miller, *Robotics: Applications and Social Implications*. Cambridge, Mass.: Ballinger, 1985.

Critchlow, Arthur J., *Introduction to Robotics*. New York: Macmillan, 1985.

Engelberger, Joseph F., *Robotics in Practice*. New York: American Management Associations, 1980.

Johnsen, Edwin G., and William R. Corliss, *Human Factors: Applications in Teleoperator Design and Operation*. New York: John Wiley & Sons, 1971.

Kapandji, I. A., *Upper Limb*. Vol. 1 of *The Physiology of the Joints*. New York: Churchill Livingstone, 1982.

L'Hote, Francois, et al., *Robot Components and Systems*. Vol. 4 of *Robot Technology*. Englewood Cliffs, N.J.: Prentice-Hall, 1983.

Logsdon, Tom, *The Robot Revolution*. New York: Simon & Schuster, 1984.

Marsh, Peter, ed., *Robots*. New York: Crescent Books, 1985.

Miller, Richard, *Automated Guided Vehicle Systems*. Fort Lee, N.J.: Technical Insights, 1983.

O'Neill, Catherine, *Computers: Those Amazing Machines*. Washington, D.C.: The National Geographic Society, 1985.

Potter, Tony, and Ivor Guild, *Robotics*. London: Usborne, 1983.

Raibert, Marc H., *Legged Robots That Balance*. Cambridge: The MIT Press, 1986.

Rehg, James A., *Introduction to Robotics: A Systems Approach*. Englewood Cliffs, N.J.: Prentice-Hall, 1985.

Robillard, Mark J., *Microprocessor Based Robotics*. New York: Howard W. Sams, 1983.

Rose, J., ed., *Survey of Cybernetics: A Tribute to Dr. Norbert Wiener*. London: Iliffe Books, 1969.

Ryan, Peter, and Ludek Pesek, *Solar System*. New York: Viking Press, 1979.

Smith, Donald N., and Peter Heytler Jr., eds., *Industrial Robots: Forecast and Trends*. Dearborn, Mich.: Society of Manufacturing Engineers, 1985.

Susnjara, Ken, *A Manager's Guide to Industrial Robots*. Englewood Cliffs, N.J.: Prentice-Hall, 1982.

Thring, M. W., *Robots and Telechirs*. New York: John Wiley & Sons, 1983.

Todd, D. J.:
Fundamentals of Robot Technology. New York: John Wiley & Sons, 1986.
Walking Machines: An Introduction to Legged Locomotion. New York: Chapman and Hall, 1986.

Vertut, Jean, and Philippe Coiffet:
Teleoperation and Robotics: Vols. 3A and 3B of Robot Technology. Englewood Cliffs, N.J.: Prentice-Hall, 1986.

Wiener, Norbert, *I Am a Mathematician*. Cambridge: The MIT Press, 1956.

Periodicals

Allan, Roger, "Nonvision Sensors." *Electronic Design*, June 27, 1985.

"Automatic Control." *Scientific American*, Sept. 1952.

Ayres, Robert, and Steve Miller, "Industrial Robots on the Line." *Technology Review*, May/June 1982.

Blakeslee, Sandra, "A Robot Arm Assists in 3 Brain Operations." *The New York Times*, June 25, 1985.

Bortz, Alfred B.:
"Joseph Engelberger: The Father of Industrial Robots Reflects on His Progeny." *Robotics Age*, Apr. 1985.
"Moravec's Mobile Robots." *Robotics Age*, Sept. 1984.

Brandt, Richard, "High Tech to the Rescue." *Business Week*, June 16, 1986.

Britton, Peter, "The New Generation of Brainy Robot Subs." *Popular Science*, Dec. 1983.

Brody, Herb, "Hands across Japan." *High Technology*, Aug. 1986.

Browne, Malcolm W., "Robot with Laser Eyes Takes Lumbering First Steps."

The New York Times, Oct. 15, 1985.

Bylinsky, Gene, "A Breakthrough in Automating the Assembly Line." *Fortune*, May 26, 1986.

Casasent, David, "Pattern Recognition: A Review." *IEEE Spectrum*, Mar. 1981.

Colligan, Douglas, "Robotic Soul." *OMNI*, June 1985.

Cooke, Patrick, "The Man Who Launched a Dinosaur." *Science 86*, Apr. 1986.

Cruickshank, Andrew, "Robots Take a Hand in Inspection, Maintenance and Repair." *Nuclear Engineering International*, Apr. 1985.

"The Cyberneticist." *Newsweek*, Mar. 30, 1964.

Dario, Paolo, and Danilo De Rossi, "Tactile Sensors and the Gripping Challenge." *IEEE Spectrum*, Aug. 1985.

"Disc Construction Allows Seven Degrees of Freedom." *Robotics World*, Apr. 1985.

Doyle, Peter C., "Bar Codes Keep Factories on Track." *High Technology*, July 1986.

Edson, Daniel, "Bin-Picking Robots Punch In." *High Technology*, June 1984.

"The Falklands: Two Hollow Victories at Sea." *Time*, May 17, 1982.

Fearing, Ronald S., and John M. Hollerbach, "Basic Solid Mechanics for Tactile Sensing." *International Journal of Robotics Research*, autumn 1985.

Feder, Barnaby J., "Factory Machines That 'See.' " *The New York Times*, Mar. 4, 1986.

"Flexible Robot Designed for Production Painting." *Robotics World*, Dec. 1984.

Friedrich, Otto, "The Robot Revolution." *Time*, Dec. 8, 1980.

Goldstein, Kenneth K., "Joseph Engelberger: A Scientist and Entrepreneur with a Special Passion for Robots." *Geo*, Dec. 1982.

Graff, Gordon, "Piezopolymers: Good Vibration." *High Technology*, June 1986.

Harmon, Leon, "Automated Tactile Sensing." *International Journal of Robotics Research*, summer 1982.

Heer, Ewald, "Robots in Modern Industry." *Astronautics & Aeronautics*, Sept. 1981.

Higgins, W.H.C., et al., "Defense Research at Bell Laboratories." *Annals of the History of Computing*, July 1982.

Hindin, Harvey J., "Algorithms Still Key to Computer Vision Systems." *Computer Design*, May 1985.

"His Master's (Digital) Voice." *Time*, Apr. 1, 1985.

Horgan, John, "Roboticists Aim to Ape Nature." *IEEE Spectrum*, Feb. 1986.

"Interview: Robert Ballard." *OMNI*, July 1986.

"J.J. Tours the *Titanic*." *Time*, July 28, 1986.

"Joseph F. Engelberger." *American Machinist*, Dec. 1982.

Kinnucan, Paul, "Robot Hand Approaches Human Dexterity." *High Technology*, Mar. 1985.

Levinson, Stephen E., and Mark Y. Liberman, "Speech Recognition by Computer." *Scientific American*, Apr. 1981.

Lovell, Clarence A., "Continuous Electrical Computation." *Bell Laboratories Record*, Mar. 1947.

Manchester, Harland, "The Electric Eye: Jack of All Trades." *The Atlantic Monthly*, Dec. 1940.

Minsky, Marvin L., "Artificial Intelligence." *Scientific American*, Sept. 1966.

Moravec, Hans P., "The Stanford Cart and the CMU Rover." *Proceedings of the IEEE*, July 1983.

Nag, Amal, "Tricky Technology: Auto Makers Discover 'Factory of the Future' Is Headache Just Now." *Wall Street Journal*, May 13, 1986.

"Norbert Wiener, 1894-1964." *Bulletin of the American Mathematical Society*. Vol. 72, No. 1, Part 2, 1966.

"A Nuclear Nightmare." *Time*, Apr. 9, 1979.

Ogorek, Michael, "Tactile Sensors." *Manufacturing Engineering*, Feb. 1985.

"On the Way to Jupiter." *Astronomy*, Jan. 1983.

Ozguner, F., et al., "An Approach to the Use of Terrain-Preview Information in Rough-Terrain Locomotion by a Hexapod Walking Machine." *Interna-*

tional Journal of Robotics Research, summer 1984.

Parrish, Michael, "The Big Guy's Back, Flying High Again and Creating a Flap." *Smithsonian,* Mar. 1986.

Pennywitt, Kirk E., "Robotic Tactile Sensing." *BYTE,* Jan. 1986.

Petre, Peter, "Speak, Master: Typewriters That Take Dictation." *Fortune,* Jan. 7, 1986.

Poe, Robert, "Where People Have a Place — For Now." *High Technology,* Aug. 1986.

Raibert, Marc H., and Ivan E. Sutherland, "Machines That Walk." *Scientific American,* Jan. 1983.

Raibert, Marc H., et al., "Experiments in Balance with a 3D One-Legged Hopping Machine." *International Journal of Robotics Research,* summer 1984.

"Right-Hand Robot." *Health,* Sept. 1985.

"Robotics in Metalworking: Engelberger on Robots." *Tooling & Production,* Oct. 1982.

"Robots Invade the Farm." *Science Digest,* May 1984.

"ROVing the Ocean Floor." *Newsweek,* Sept. 26, 1983.

Saveriano, Jerry W., "An Interview with Victor Scheinman." *Robotics Age,* autumn 1980.

"Scientists See Technological Advances from *Titanic* Mission." *Santa Cruz Sentinel,* July 27, 1986.

Seering, Warren P., "Who Said Robots Should Work Like People?" *Technology Review,* Apr. 1985.

"Sharpshooting at Unseen Targets." *Popular Mechanics,* Dec. 1939.

Solomon, Stephen, "Miracle Workers." *Science Digest,* Dec. 1981.

Thompson, Alan, "Introduction to Robot Vision." *Robotics Age,* summer 1979.

Tierney, John, "Yik San Kwoh Built a Better Brain Surgeon." *Esquire,* Dec. 1985.

Tucker, Jonathan B., "Submersibles Reach New Depths." *High Technology,* Feb. 1986.

Wiener, Norbert, "Cybernetics." *Scientific American,* Nov. 1948.

Yeates, Clayne M., and Theodore C. Clarke, "Galileo: Mission to Jupiter." *Astronomy,* Feb. 1982.

Zimmerman, Marlene, "The Robot Master." *Datamation,* Apr. 1982.

Other Sources

"The Autonomous Land Vehicle Program." Denver, Colo.: Martin Marietta, Dec. 1985.

Bares, John, Lee Champeny and W. L. Whittaker, *Three Remote Systems for TMI-2 Basement Recovery.* Pittsburgh, Pa.: Carnegie Mellon, no date.

Bolles, Robert C., and Patrice Horaud, "3DPO: A Three-Dimensional Part Orientation System." Submitted to the *International Journal of Robotics Research,* Feb. 1985.

Dario, Paolo, et al., "Tendon Actuated Exploratory Finger with Polymeric, Skin-Like Tactile Sensor." *1985 IEEE Conference on Robotics and Automation.* Silver Spring, Md.: IEEE Computer Society Press, 1985.

Fisher, Scott S., "Virtual Interface Environment." Moffett Field, Calif.: NASA Ames Research, July 1986.

Hightower, J. D., and D. C. Smith, "Teleoperator Technology Development." Kailua, Hawaii: Naval Ocean Systems Center, no date.

Holbrook, Bernard D., and W. Stanley Brown, "A History of Computing Research at Bell Laboratories (1937-1975)." *Computing Science Technical Report No. 99.* Bell Telephone Laboratories, Mar. 1982.

Hollerbach, John M., "Robot Hands and Tactile Sensing." M.I.T. Artificial Intelligence Laboratory, Jan. 1986.

Laeser, Richard P., "Engineering the Voyager Uranus Mission." *IEEE EasCon 86, 19th Annual Electronics and Aerospace Systems Conference,* Sept. 8-10, 1986.

Lasecki, Robert R., "AGV System Criteria." *Robots 10 Conference Proceedings.* Dearborn, Mich.: Robotics International of SME, 1986.

Lyons, Damian M., "A Simple Set of Grasps for a Dextrous Hand." *1985 IEEE Conference on Robotics and Automation.* Silver Spring, Md.: IEEE Computer Society Press, 1985.

McGhee, Robert B., "Robot Locomotion." *Neural Control of Locomotion.* New York: Plenum, 1976.

McGhee, Robert, et al., "A Hierarchically Structured System for Computer Control of a Hexapod Walking Machine." *Theory and Practice of Robots and Manipulators.* Cambridge: The MIT Press, 1985.

Moravec, Hans P., "The Cart Project: A Personal History, a Plea for Help and a Proposal." May 1974.

"Nuclear Safety after TMI." *EPRI JOURNAL,* June 1980.

Robotic Industries Association, *RIA Robotics Glossary.* Dearborn, Mich.: Robot Institute of America, 1984.

Thorpe, Charles E., "Robotic Vehicles." Pittsburgh, Pa.: Carnegie Mellon Robotics Institute, June 1986.

Tracey, P. M., "Automated Guided Vehicle Safety." *Robots 10 Conference Proceedings.* Dearborn, Mich.: Robotics International of SME, 1986.

Venturini, Marco, "Brushless Drives." Paper presented at Giornata di Studeio del 7/6/84 organizzata dalla Sezione di Milano dell'AEI, July 1984.

"Voyager Nears Uranus." Press release 85-165. Washington, D.C.: National Aeronautics and Space Administration.

Wilson, Jack, and Doug MacDonald, "Telepresence: Goal or By-Product of Remote Systems." *Robots 10 Conference Proceedings.* Dearborn, Mich.: Robotics International of SME, 1986.

Wilson, James F.:
Compliant Robotic Structures: First Report to DARPA, June 1, 1984-May 31, 1985. Springfield, Va.: National Technical Information Service, Aug. 1985.
Compliant Robotic Structures — Part II: Second Report to DARPA, June 1, 1985-May 31, 1986. Alexandria, Va.: Defense Technical Information Center, July 1986.

Zue, Victor W., "Speech Recognition by Machine." M.I.T. Department of Electrical Engineering and Computer Science, Jan. 1986.

Acknowledgments

The index for this book was prepared by Mel Ingber. The editors also thank: **In France:** Cachan — Arnauld Laffaille, Association Française De Robotique Industrielle; Parley 2 — Jean-Claude Goaër, Renault Automation; Vanves — Philippe Louste, Matra Transports. **In Japan:** Fukuoka — Hisashi Kojima, Kaho Musen Co. Ltd.; Oshi — Masayo Matsukawa, Fanuc LTD; Tokyo — Yoshiyuki Matsumoto, Tomy Company Inc.; Youichi Suga, Hitachi, Ltd. **In Sweden:** Mölndal — Ove Larson, Spine Robotics AB. **In the United States:** California — Menlo Park: Robert Bolles, SRI International; Moffett Field: David Ennis, Scott Fisher, Henry Lum, NASA Ames Research Center; Pasadena: Richard Laeser, Donna Wolff, Jet Propulsion Laboratory; Carver Mead, California Institute of Technology; Petaluma: Jim Dietrick, Compumotor Corporation; Rohnert Park: Michael Chung, Parker Hannifin Corporation; San Diego: Mark Raptis, General Dynamics; Connecticut —Bethel: Joseph Engelberger, Transitions Research Corp.; District of Columbia — Lauren J. Barth, National Zoo; Dana Bell, National Air and Space Museum; Clare Miles, Nuclear Regulatory Commission; Georgia — Atlanta: Kirk Pennywitt, Georgia Tech Research Institute; Kentucky — Lexington: Tim Holt, IBM; Maryland — Gaithersburg: James Albus, Alexander Slocum, National Institute of Standards and Technology; Upper Marlboro: Roy Truman, Eastport International; Massachusetts — Amherst: Allen R. Hanson, University of Massachusetts at Amherst; Cambridge: Lori Lamel, John Hollerbach, Kathryn Kline, Walter Rosenblith, J. K. Salisbury, Heidi Saraceno, Massachusetts Institute of Technology; Daniel Whitney, Charles Stark Draper Laboratories; Michigan — Grand Rapids: Kenneth Ruehrdanz, Barrett Vehicle Systems; Warren: Richard Beecher, Anthony Gagliardi, Gerard Graye, General Motors; Missouri — St. Louis: Susan Flowers, Katherine MacDonald, McDonnell Douglas Astronautics Company; New Hampshire — Amherst: David Vander Mey, Vanguard Technology Ventures; New Jersey — Short Hills: William Bucci, M. D. Fagen, Michael Jacobs, Bell Laboratories; New York — Cornwall-on-Hudson: John R. Elting; North Carolina — Durham: James Wilson, Duke University; Ohio — Toledo: Gail Christie, Dan Hasselschwert, The DeVilbiss Company; Oregon — Corvalis: Bruce Paris, Intelledex Corp.; Pennsylvania — Bethlehem: Mikkell Groover, Lehigh University; Middletown: Debbie Phillips, Gordon Tomb, Three Mile Island; Pittsburgh: Richard Stern, Hans Moravec, Friedrich Prinz, Todd Simonds, William L. Whittaker, Carnegie Mellon; Utah — Salt Lake City: Barry Hanover, Animate Systems Incorporated. Virginia — Merrifield: Frances Ford, James Haddock, Gerald Merna, Horace Oliver, Sandra Stewart, Merrifield Post Office; Wisconsin — Milwaukee: Richard Braatz, John Rothwell, Allen-Bradley Corp. **In West Germany:** Landsberg — Wolfgang Klinker, *Roboter* magazine.

Picture Credits

The sources for the illustrations that appear in this book are listed below. Credits from left to right are separated by semicolons; credits from top to bottom are separated by dashes.
Cover, 6: Art by Nick Gaetano/Kahn. 12, 13: Art by Matt McMullen. 19-31: Art by Stephen Wagner. 32: Art by Nick Gaetano/Kahn. 34-45: Art by Matt McMullen. 47-59: Art by Greg Harlin/Stansbury, Ronsaville, Wood Inc. 60: Art by Nick Gaetano/Kahn. 64-67: Art by John Drummond. 71-73: Art by Peter Sawyer. 76, 77: Courtesy The Robotics Institute of Carnegie Mellon; courtesy Martin Marietta; Kevin Fitzsimmons/NYT Pictures — Dan McCoy/Rainbow; © 1984 Marc H. Raibert; Hans Moravec, Stanford University; courtesy SRI International — Science Photo Library, London; Hitachi, Ltd., Tokyo; Mike Milocek, courtesy Center for Engineering Design, University of Utah; courtesy Dr. Shigeo Hirose, Tokyo Institute of Technology. Background art by John Drummond. 81-89: Art by Stephen Wagner. 90-99: Art by Steve Bauer/Bill Burrows Studio. 100: Art by Nick Gaetano/Kahn. 107-109: Art by Peter Sawyer. 115-121: Art by Marvin Fryer.

Index

Time-Life Books is a division of Time Life Inc.,
a wholly owned subsidiary of
THE TIME INC. BOOK COMPANY

TIME-LIFE BOOKS

MANAGING EDITOR: Thomas H. Flaherty
Director of Editorial Resources:
Elise D. Ritter-Clough
Director of Photography and Research:
John Conrad Weiser
Editorial Board: Dale M. Brown, Roberta Conlan,
Laura Foreman, Lee Hassig, Jim Hicks, Blaine Marshall, Rita Thievon Mullin, Henry Woodhead

PUBLISHER: Joseph J. Ward

Associate Publisher: Trevor Lunn
Editorial Director: Donia Ann Steele
Marketing Director: Regina Hall
Director of Design: Louis Klein
Production Manager: Marlene Zack
Supervisor of Quality Control: James King

Editorial Operations
Production: Celia Beattie
Library: Louise D. Forstall
Computer Composition: Deborah G. Tait (Manager),
Monika D. Thayer, Janet Barnes Syring, Lillian Daniels

Correspondents: Elisabeth Kraemer-Singh (Bonn);
Maria Vincenza Aloisi (Paris); Ann Natanson (Rome).
Valuable assistance was also provided by: Wanda
Menke-Glückert (Bonn); Christine Hinze (London);
John Dunn (Melbourne); Elizabeth Brown and Christina Lieberman (New York); Ann Wise (Rome); Mary
Johnson (Stockholm); Dick Berry and Mieko Ikeda
(Tokyo).

UNDERSTANDING COMPUTERS

SERIES DIRECTOR: Roberta Conlan
Series Administrator: Loretta Britten

Editorial Staff for *Robotics*
Designer: Ellen Robling
Associate Editors: Sara Schneidman (pictures)
Researchers: *Text Editors:*
Elisabeth Carpenter   Allan Fallow
Esther Ferington   Ray Jones
Flora Garcia   Peter Pocock
Assistant Designers: Antonio Alcalá, Sue Deal,
Christopher M. Register
Editorial Assistant: Miriam Newton Morrison
Copy Coordinators: Vilasini Balakrishnan,
Robert M. S. Somerville
Picture Coordinators: Renée DeSandies,
Bradley Hower

Special Contributors: Ronald H. Bailey, Avery
Comarow, Donal Kevin Gordon, Mark A. Hofmann,
Robert W. Hunnicutt, Carollyn James, Martin Mann,
Valerie Moolman, Peter Radetsky, Jeffrey Rothfeder,
Charles C. Smith, William Worsley (text); Roxie
France-Nuriddin, Richard A. Davis (research)

CONSULTANTS

HOWARD M. BLOOM is chief of the Factory Automated
Systems Division at the National Bureau of Standards,
where he is responsible for designing systems for an automated manufacturing research facility.

DANILO EMILIO DE ROSSI is a research associate at
the University of Pisa and a visiting professor in bioengineering at Brown University. He has recently worked
on the development of artificial organs and sensors.

GEORGE C. DEVOL JR., often called the father of the
industrial robot, has pioneered in the electronics industry
as an inventor and entrepreneur. He is currently the chairman of the board of Automatic Manufacturing Systems.

WILLIAM R. HAMEL is with the Instrumentation and
Controls Division at the Oak Ridge National Laboratory in Tennessee, where he directs the development of
robotics-related technology for teleoperations in radioactive environments.

SHIGEO HIROSE is an associate professor at the Tokyo
Institute of Technology, where he studies mechanisms,
sensors and control of robots.

ROBERT LASECKI, vice president of The Austin Company, a software-manufacturing consulting firm in Rosemont, Illinois, has been working in the field of automated
guided vehicles since 1970.

ROBERT B. MCGHEE, who designed a number of walking machines while at Ohio State University, is currently
with the Department of Computer Science at the Naval
Postgraduate School in Monterey, California.

NICHOLAS ODREY is director of the Robotics Laboratory
at Lehigh University. He is also chairman of the Research
and Development Division of Robotics International of
the Society of Manufacturing Engineers.

ISABEL LIDA NIRENBERG has dealt with a wide range
of computer applications, from the analysis of data collected by the Pioneer space probes to the matching of
children and families for adoption agencies. She works
in the Computer Center of the State University of New
York at Albany.

TED PANOFSKY, formerly with the Stanford University
Artificial Intelligence Laboratory, is the manager of Advanced Projects at Machine Intelligence Corporation,
manufacturers of advanced automated inspection and
factory automation in Sunnyvale, California.

MARC H. RAIBERT is a member of the Artificial Intelligence Laboratory at the Massachusetts Institute of
Technology, where he is an associate professor of electrical engineering and computer science. His research
deals with the interactions between computers and the
physical world.

WARREN P. SEERING is associate professor of mechanical engineering at the Massachusetts Institute of Technology. A member of M.I.T.'s Artificial Intelligence Laboratory, he is working on the design of high-performance
robotics systems.

CHARLES M. STRACK is production marketing manager
for Omron Electronics Inc. in Schaumburg, Illinois. He
has extensive experience in the application of industrial
sensor technology.

CHARLES E. THORPE is a research scientist at the Robotics Institute of Carnegie Mellon, with a special interest in
vision and navigation for mobile robots.

GABRIELE VASSURA is an associate professor of mechanical engineering at Bologna University, where he
has concentrated on the design of automatic machinery.
His most recent work has been in the making and application of a dexterous, articulated hand.

RICHARD WISEMAN is director of Robotics and Advanced Controls at Foster Miller Inc. in Waltham, Massachusetts. He received his Ph.D. in mechanical engineering from the Massachusetts Institute of Technology.

VICTOR W. ZUE is an assistant professor of electrical
engineering and computer science at the Massachusetts
Institute of Technology, where his primary research is in
the application of acoustic phonetic knowledge to the
development of speech-recognition systems.

Library of Congress Cataloging-in-Publication Data
Robotics / by the editors of Time-Life Books.
    p. cm. — (Understanding computers)
  Includes bibliographical references and index.
  ISBN 0-8094-7582-0 — ISBN 0-8094-7583-9 (lib. bdg.)
  1. Robotics.   I. Time-Life Books.   II. Series.
TJ211.15.R62 1991
629.8'92—dc20   91-16459
                                                      CIP

For information on and a full description of any of the Time-
Life Books series listed, please write:
Reader Information
Time-Life Customer Service
P.O. Box C-32068
Richmond, Virginia 23261-2068

REVISIONS STAFF

EDITOR: Lee Hassig

Writer: Esther Ferington
Assistant Designer: Bill McKenney
Copy Coordinator: Jill Lai Miller
Picture Coordinator: Katherine Griffin

Consultants: James Parker, Charles E. Thorpe